Women's Food Matters

"This groundbreaking interdisciplinary feminist study offers a new perspective on how, and why, women's food *matters* throughout history and in our contemporary world. As one of the first studies to combine a focus on food production, processing and cooking, on food cultures and food systems, Swinbank puts women's knowledge and creativity at center stage in the reproduction and transformation of culture and agriculture. *Women's Food Matters* provides a theoretically rich contribution that is jargon-free, making it an appropriate choice for classes at any level, as well as for the general reader. Destined to 'stir the pot' of contemporary food studies."

—David E. Sutton, *Professor of Anthropology at Southern Illinois University*

"*Women's Food Matters* provides a comprehensive and overarching historical and cross-cultural view of women's food-work and the important role women have played in shaping the food landscape from production through consumption. Drawing from examples around the world, Swinbank illustrates how women's food knowledge and practices must be considered in addressing some of the most pressing problems facing the food system today."

—Deborah Harris, *Associate Professor of Sociology at Texas State University, San Marcos, Texas*

"Radical feminism has just taken its long-awaited seat at the food studies table. In *Women's Food Matters*, Vicki Swinbank reminds us that women's intergenerational food knowledge – its production, preparation and consumption - is at the heart of most food cultures. In her original radical feminist analysis of women's role in various food systems throughout history, Swinbank powerfully sets out the erosion of women's knowledge by patriarchal and capitalist systems that have contributed to everything from taking credit for women's recipes in contemporary culinary culture to industrialised farming and genetically-modified crops. *Women's Food Matters* is consistently engaging, informative and persuasively argued; both taking us back to the wonderful memories of being in grandma's kitchen, and into the diverse and widely-politicised world of the global food system."

—Natalie Jovanovski, *Lecturer and DECRA Research Fellow, The University of Melbourne, Australia*

Vicki A. Swinbank

Women's Food Matters

Stirring the Pot

palgrave
macmillan

Vicki A. Swinbank
Northcote, VIC, Australia

ISBN 978-3-030-70395-0 ISBN 978-3-030-70396-7 (eBook)
https://doi.org/10.1007/978-3-030-70396-7

Cover credit: imageBROKER/Alamy Stock Photo

This Palgrave Macmillan imprint is published by the registered company Springer Nature Switzerland AG
The registered company address is: Gewerbestrasse 11, 6330 Cham, Switzerland

In memory of my mother
whose love of cooking lives on in me

Acknowledgements

This book is a culmination of my two lifelong passions: food issues and cooking, and feminism. There are several people I would like to acknowledge without whose support and interest this book might not have seen the light of day.

I am indebted to Natalie Jovanovski for her encouragement and enthusiasm throughout this project, especially in the early stages, which strengthened my belief in the importance of my book. I would also like to thank Sue Leigh for her insightful and wise comments from reading some early drafts. My thanks go to Helen Chambers for her constant interest and helpful advice about the publishing process. My friends in the United Kingdom, Elaine Hutton and Lynne Harne followed the progress of the book from afar and were supportive from the start. I also appreciate the interest of people, both known and unknown, with whom I have talked about this project, and who said that they were looking forward to reading the book.

Above all, I am deeply grateful to Kathy Chambers for her invaluable technical support throughout the project as well as her editing assistance in the final stages; it would not have been possible without her. I am also grateful for her love, support and forbearance during the sometimes stressful writing process.

Finally, I wish to thank my editors at Palgrave Macmillan, firstly, Amelia Derkatsch in London and more recently, Nina Guttapalle in New York, who were always helpful and prompt with answers to my questions, as well as the rest of the team at Palgrave and at Springer Nature.

Contents

CHAPTER 1

Introduction

In recent years there has been a veritable explosion in books dealing with food issues, ranging from the history of cooking to a critique of the industrialised food system. There is, however, relatively little that deals with food from a feminist perspective, with much of the existing literature focusing on women's problematic relationship with food, especially eating disorders and body image (Bordo 2004; Jovanovski 2017; Manton 1999) and women's ambivalent relationship with cooking (DeVault 1991; Cairns and Johnston 2010). Whilst this work by feminist academics has made an invaluable contribution to the literature on women's often conflicted relationship to food, my aim in this book is to examine the more positive and empowering aspects of women's involvement in food production, preparation and cooking.

My analysis in this book is from a radical feminist standpoint, that is a politics that "places women and women's experiences at the centre, names the oppression of women, and involves a holistic view of the world, an analysis which probes every facet of existence for women" (Rowland and Klein 1996, 13). Radical feminist theory is concerned about the material reality of women's lives, recognising that the 'personal is political'—in other words, that women's personal experiences are a consequence and reflection of the power imbalances in hetero-patriarchal society resulting from a gendered hierarchy based on sexual difference. This is exacerbated and complicated by class, race and ethnicity. Radical feminists argue that

V. A. Swinbank, *Women's Food Matters*,
https://doi.org/10.1007/978-3-030-70396-7_1

the system of male domination and female subordination that gendered power relations are based upon is *not* a result of biology or essentialism, but is a *cultural construction,* used to justify the socially constructed gendered hierarchical system of masculinity and femininity.

Significantly, although there is a considerable body of literature on the history of cooking, it is overwhelmingly androcentric in its analysis, with men seen as the central players in the development of the uniquely human activity of cooking (Wrangham 2009; Standage 2009; Revel 1982; Fernandez-Armesto 2004). Even those texts written by women tend to concentrate on the cooking of famous male chefs, such as Escoffier (Trubek 2000). There is also a very limited body of literature on women's historical role as food providers, the most notable being in the classic book '*Woman the Gatherer*' (Dahlberg 1981).

In contrast to this, in Chapter 2, I apply a feminist analysis to the origins of cooking, arguing that it was almost certainly early female hominids who developed the first technology to do with cooking (Zhilman 1981; Ehrenberg 1989; Stanley 1981). Drawing on this and writings by Hawkes et al. (2002) and Gremillion (2011), I describe how women, in their historical role of provisioning and preparing food have been instrumental in the development of cooking. I apply a feminist critique of the prevailing male-centred history of cooking, which credits 'man the hunter' with the discovery of cooking with fire (Wrangham 2009; Standage 2009). Furthermore, I challenge the assumption of many, mostly male, historians (e.g. Revel 1982; Dickie 2008) that the cooking of the common people/peasantry (in other words, women's domestic cooking) was unimaginative, dull and monotonous. I contend that, on the contrary, traditional, female domestic cooking has been hugely innovative, imaginative and varied. Moreover, in their need to make the best use of what is seasonally and locally available, women's domestic cooking has, over many centuries, if not millennia, formed the basis of the world's myriad regional cuisines.

Chapter 3 explores why men's cooking—especially professional or '*haute cuisine*'—has always been awarded a higher status than women's domestic cooking. This is despite the fact that the former depends on the latter for its existence. I analyse how the elevation of men's cooking to a higher status has involved the denial of such dependence and indeed even a denigration of women's cooking (Revel 1982). The evolution of cuisines is often considered to be from the top down—that is, the influence of 'high' or 'haute' cuisine on everyday women's domestic cooking. I argue, in my feminist analysis of 'culinary hierarchy'(Swinbank 2002), that the main influence on cooking cultures is in fact the other way around—that is, 'haute' cuisine's unacknowledged appropriation of and dependence on women's traditional domestic cooking. The masculinist assertion that men are superior cooks to women continues today, perhaps even more so as cooking has, in many Western societies, become viewed as a trendy lifestyle activity taken up as a leisure activity by many men. I examine this in the following chapter.

Domestic cooking and the associated domestic kitchen have traditionally been a female sphere, one of the few areas where women legitimately have been able to exercise agency and creativity. However, with the rise of male celebrity chef culture in recent years, cooking (at least on an occasional basis) has now become a trendy activity for many men. Whilst this has been welcomed by some women, men's 'takeover' of the domestic kitchen has at times made it a contested space, adding to a sense of superiority of some men, vis-à-vis women, with regard to cooking. This is an issue I examine in Chapter 4, with reference to geography academic Angela Meah (2013), whose valuable work on this issue I draw on and extend.

This chapter explores the contemporary trend for men to become increasingly involved in the kitchen and the challenges that this presents for women. I conclude with a look at the impact of male celebrity chefs on cooking, and examine how they claim to have discovered various approaches to cooking (e.g. foraging, peasant-style food, slow cooking) that have in fact been used for centuries by women. This appropriation of women's food knowledge connects back to the main themes of the previous two chapters—namely the claim that men developed the uniquely human activity of cooking. The masculinist claim that men are innately better cooks than women is not only a falsehood, but also a source of great irritation to many women. The elderly mother-in-law of an acquaintance of mine expressed this frustration by exclaiming "Can't they even leave us that!".

It may be puzzling that, given women's central role in most aspects to do with food, not only cooking, but also, in much of the world, growing and preserving it, there has been relatively little interest in this, including (with some exceptions) by feminists. This is in part due to second-wave feminists viewing women's role in domestic cooking as a prime example of their subordination and exploitation, and therefore, to be sidelined or even dispensed with in the interests of women's liberation. As I argue in Chapter 5, this is a case of "throwing the baby out with the bathwater", and that, instead it is very much in the interests of feminism to acknowledge and celebrate the ingenuity and creativity of women's domestic cooking. This has in fact created the myriad regional cuisines of the world, and been instrumental in the development of the world's culinary cultures. Another important reason for this oversight is that, especially in modern Western culture, women's unpaid work in the domestic sphere is taken for granted and so rendered unimportant and invisible, an issue that feminists have analysed in depth (Waring 1988; Delphy and Leonard 1992; Mies et al. 1988).

The intergenerational transfer of food knowledge is fundamental to the development and maintenance of food cultures. Over many centuries/millennia, in their role as food producers, providers and preparers, women have played a central role in this. However, once again, this has been largely overlooked or ignored. In Chapter 5, I delve into the largely unexplored world of the intergenerational transfer of food knowledge between women. This female intergenerational food knowledge and the associated culinary cultural memory which forms the basis of the world's innumerable regional food cultures, has been, and in many places still is traditionally transmitted orally and by observation. Similarly, the female intergenerational handing on of handwritten cookbooks, as culinary heirlooms, are treasured for their emotional significance, in addition to being highly valued for their wealth of recipes and food knowledge. Far from being a trivial matter, this intergenerational transfer of women's food knowledge, both oral and written, is vital to the development and maintenance of the world's regional food cultures. Although food cultures are now beginning to be recognised by the United Nations as a vital aspect of human cultural heritage, women's role in this remains generally unacknowledged.

This chapter also includes a critique of feminist arguments that reject cooking as a site of female subordination. Whilst recognising that women often have an ambivalent attitude to cooking, given that it is generally not optional with regard to feeding families, I contend that to see it solely in terms of exploitation, and therefore to be abandoned, is a mistake. This is because in doing so women unwittingly reject a crucial part of women's contribution to human culture and civilisation—in short, a proud female heritage. It also allows men to claim this heritage for themselves, as I have analysed in previous chapters.

In this book, I use the term 'masculinist' to describe how in male supremacist culture the gendered conditioning of men into masculinity results in attitudes of dominance and a sense of entitlement. The consequences of this are the cause both of the cultural subordination and oppression of women as well as of an exploitative, extractive attitude to nature. This is accentuated by capitalism, itself a product of masculinist mindset. Chapter 6 describes in particular how the development of capitalist agriculture has wrought untold damage on the environment, including being a major cause of climate change. Much has been written about capitalism and its origins, most famously of course by Karl Marx; however, there is little that deals with the historical effect of capitalism on women, and even less—with some important exceptions (Humphries 1990; Sharpe 2016)—dealing with the impact on women's traditional role in subsistence food production and preparation. In this chapter, I examine in some detail, starting with the enclosure of the commons in England, the destructive effects on women of the development of capitalist industrial food systems. The impact on women's traditional role as food producers, and the accompanying loss of related skills and status were fundamentally altered/destroyed by the development of the capitalist industrial food system, a process that started in the West, specifically historically in England, but which now threatens much of the world. Both industrial food systems (exemplified by monocultures, genetically modified crops and factory farming of animals) and '*haute cuisine*' at the other end of the food spectrum (exemplified by molecular gastronomy) are products of a masculinist mindset.

Chapter 6 starts with an analysis of the damaging effects of the enclosure of the commons and the Industrial Revolution on British (more specifically English) food culture. The enclosures, which preceded and accompanied the industrialisation of Britain, dispossessed the peasantry of their land and livelihoods, turning them into an urban proletariat workforce for the newly emerging industries. The loss of land from which to grow, gather and feed themselves had a devastating effect on the dispossessed rural population and a previously vibrant food culture. Crucially these developments resulted in the loss of women's food knowledge and skills, as well as having negative consequences for their economic status. In the second part of the chapter, I look at the impact of the Scientific Revolution on agriculture and food production, leading to a mechanistic view of nature (Merchant 1989). This profound shift in thinking, along with the Acts of Enclosure, laid the foundation for the development of capitalism and the eventual full industrialisation of the food system which exists in much of the world today. I examine the development in the twentieth century of chemical farming and the introduction of genetically modified crops, and the harm caused by this to the health of people, animals and the environment. This destructive and unsustainable food system is now being increasingly imposed on the Global South, with serious negative effects on female peasant farmers who traditionally produce the majority of food in an environmentally sound way.

In Chapter 7, I describe the threats posed to food security and food sovereignty, in which women play a central role. These threats include industrialised food systems, the use of toxic chemicals in agriculture and the development of genetically modified crops, which pose a serious threat to biodiversity. I also examine how the dependence of industrialised food systems on fossil fuels is a major contributor to climate change, and hence a very serious threat to food security. I argue that the contrasting biodiverse, ecologically sustainable practices of the world's female peasant farmers provide a solution to the damaging effects of industrial farming, including that of climate change.

Throughout the world, women in their role as food producers, food processors, providers and preparers for their families have traditionally been key to the development and maintenance of the world's food systems and cultures. Again this has not been sufficiently recognised/acknowledged. One striking example of this lack is that of British writer on agricultural issues, Colin Tudge (2004) who says an insight or 'revelation' prompted him to write a book on the relationship between sustainable farming systems and good cooking and nutrition: "It came to me as if on high that all the world's great cooking – Chinese, Indian, peasant French and Italian, Turkish, north African, South-East Asian, East European – is based on products of traditional farming, for how could it be otherwise? In other words, … great cooking has evolved over hundreds or indeed thousands of years to make the best use of wild nature and traditional farming" (Tudge 2004, 6). However, despite this important 'revelation', Tudge makes absolutely no mention of women in his book, despite women's crucial and central historical role in food systems, both production and preparation, throughout the world, a role that has almost certainly made them largely responsible for the development of sustainable food systems and the related cuisines mentioned by Tudge. Tudge is certainly not alone in this omission; a glance at the index of most books dealing the history of food and agricultural food systems reveals a glaring absence of any mention of women. Even, the esteemed female French historian, Maguelonne Toussaint-Samat's encyclopaedic 'History of Food' (1987) makes only one mention of women and that is to do with the prohibition on women of drinking wine in ancient Rome.

In Chapter 8, I attempt to address this serious oversight by discussing women's crucial role in the vitally important relationship between biodiversity and culinary diversity, a relationship, which supports ecologically sustainable food systems as well as food cultures. The intimate connection between biodiversity and cultural culinary diversity and women's central role in this is only just beginning to be recognised, albeit by a very few, mostly female ethnobotanists (Howard 2003; Greenberg 2003; Hoffman 2003; Ertug 2003). In this chapter, I examine women's traditional role in a wide range of activities associated with biodiversity and culinary diversity: women's homegardens, which represent a rich source of biodiversity; women's role as food-gatherers of wild plants, and their culinary application, as well as their medicinal use; women as the custodians of seeds; the many ingenious ways women have developed over the centuries of processing and preserving food and women's work with domesticated food-producing animals (Sachs 1996). Recognition of women's central role in sustainable systems of food production, what is now referred to as food sovereignty, is of vital importance because it provides not only a contrast/alternative to, but, even more importantly, provides the solution to the current environmental and health crises caused by industrialised food systems.

In *Women's Food Matters* I emphasise that women's crucial role in developing and maintaining sustainable food systems, including their central role in the development of the world's countless regional food cultures, has not been sufficiently recognised, and indeed is generally entirely overlooked and even denied. Redressing this omission thus is a central theme of this book. Through the interdisciplinary and feminist analysis presented in this book, the central role of women in the development and preservation of food cultures is fully acknowledged perhaps for the first time.

References

Bordo, Susan. 2004. *Unbearable Weight: Feminism, Western Culture, and the Body.* Berkeley: University of California Press.

Cairns, Kate, Josee Johnston, and Shyon Baumann. 2010. "Doing Gender in the Foodie Kitchen." *Gender and Society* 24 (5): 591–615.

Dahlberg, Frances, ed. 1981. *Woman the Gatherer*. New Haven and London: Yale University Press.

Delphy, Christine, and Diana Leonard. 1992. *Familiar Exploitation: A New Analysis of Marriage in Contemporary Western Societies*. Cambridge: Polity Press.

DeVault, Marjorie L. 1991. *Feeding the Family: The Social Organization of Caring as Gendered Work*. Chicago and London: The University of Chicago Press.

Dickie, John. 2008. *Delizia: The Epic History of the Italians and Their Food*. New York and London: Free Press.

Ehrenberg, Margaret. 1989. *Women in Prehistory*. London: British Museum Publications.

Ertug, Fusun. 2003. "Gendering the Tradition of Plant Gathering in Central Anatolia (Turkey)." In *Women and Plants: Gender Relations in Biodiversity Management and Conservation*, edited by Patricia L. Howard. 183–196. London and New York: Zed Books.

Fernandez-Armesto, Felipe. 2004. *Near a Thousand Tables: A History of Food*. New York: Free Press.

Greenberg, Laurie, S. Z. 2003. "Women in the Garden and Kitchen: The Role of Cuisine in the Conservation of Traditional House Lot Crops among Yucatec Mayan Immigrants." In *Women and Plants: Gender Relations in Biodiversity Management and Conservation*, edited by Patricia L. Howard, 51–62. London and New York: Zed Books.

Gremillion, Kristen, J. 2011. *Ancestral Appetites: Food in Prehistory*. Cambridge: Cambridge University Press.

Hawkes, K., J. F. O'Connell, N. G. Blurton Jones, H. Avarez, and E. L. Charnov. 2002 'The Grandmother Hypothesis and Human Evolution'. In *Adaptation and Human Behavior: An Anthropological Perspective*, edited by L. Cronk, N. Chagnon and W. Irons, 231–252. New York: Aldine de Gruyter.

Hoffman, Shirley. 2003 "Arawakan Women and the Erosion of Traditional Food Production in Amazonas Venezuela." In *Women and Plants: Gender Relations in Biodiversity Management and Conservation*, edited by Patricia L. Howard, 258–269. London and New York: Zed Books.

Howard, Patricia, L. 2003 "Women and the Plant World: An Exploration." In *Women and Plants: Gender Relations in Biodiversity Management and Conservation*, edited by Patricia L. Howard, 1–48, London and New York: Zed Books.

Humphries, Jane. 1990. "Enclosures, Common Rights, and Women: The Proleterianization of Families in the Late Eighteenth And Early Nineteenth Centuries." *The Journal of Economic History* (1): 17–42.

Jovanovski, Natalie 2017. *Digesting Femininities: The Feminist Politics of Contemporary Food Culture*. Cham, Switzerland: Palgrave Macmillan, Springer International Publishing.

Manton, Catherine 1999. *Fed Up: Women and Food in America*. Westport, CT and London: Bergin & Garvey.

Meah, Angela, and Peter Jackson. 2013 *Crowded Kitchens: The Democratisation of Domesticity*. Gender, Place and Culture. https://doi.org/10.1080/0966396x.2012.701202.

Merchant, Carolyn. 1989. *The Death of Nature: Women, Ecology and the Scientific Revolution*. New York: HarperCollins.

Mies, Maria, Veronica Beenholdt-Thomsen, and Claudia von Werlhof. 1988. *Woman, the Last Colony*. London: Zed Books.

Revel, Jean-Francois. 1982. *Culture and Cuisine: A Journey Through the History of Food*. Translated by Helen R. Lane. New York: Da Capo.

Rowland, Robyn, and Renate Klein. 1996. "Radical Feminism: History, Politics, Action." In *Radically Speaking: Feminism Reclaimed*, edited by Diane Bell and Renate Klein, 9–36. Melbourne: Spinifex Press.

Sachs, Carolyn. 1996. *Gendered Fields: Rural Women, Agriculture, and the Environment*. Boulder, CO: Westview Press.

Sharpe, Pamela. 2016. *Adapting to Capitalism: Working Women in the English Economy, 1700–1850*. Basingstoke, UK: Palgrave Macmillan.

Standage, Tom. 2009. *An Edible History of Humanity*. New York: Walker Publishing.

Stanley, Autumn. 1981. "Daughters of Isis, Daughters of Demeter: When Women Sowed and Reaped." *Women's Studies International Quarterly* 4 (3): 289–304.

Swinbank, Vicki A. 2002. The Sexual Politics of Cooking: A Feminist Analysis of Culinary Hierarchy in Western Culture. *Journal of Historical Sociology* 15 (4): 464–494.

Toussaint-Samat, Maguelonne. 2009. *History of Food*. Translated from the French by Anthea Bell. Chichester, West Sussex, UK and Malden, MA: Wiley-Blackwell.

Trubek, Amy. 2000. *Haute Cuisine: How the French Invented the Culinary Profession*. Philadelphia: University of Pennsylvania Press.

Tudge, Colin. 2004. *So Shall We Reap: What's Gone Wrong with the World's Food—And How to Fix it*. London: Penguin.

Waring, Marilyn. 1988. *Counting for Nothing: What Men Value and What Women Are Worth*. Wellington: Allen and Unwin.

Wrangham, Richard. 2009. *Catching Fire: How Cooking Made Us Human*. London: Profile Books.

Zhilman, Adrienne, L. 1981. "Women as Shapers of Human Adaptation." In W*oman the Gatherer*, edited by Frances Dahlberg. New Haven and London: Yale University Press.

CHAPTER 2

A Feminist History of Cooking: From Hunter-Gatherer to Peasant Cooking

French anthropologist, Claude Levi-Strauss considered cooking—which, like language, is a "truly universal form of human activity"—to be the key human activity that defines the origins of human culture. Indeed, according to him, "not only does cooking mark the transition from nature to culture, but through it and by means of it, the human state can be defined in all its attributes" (cited in Symons 2004, 103). Other anthropologists consider that the conversion of raw material into cooked food with the use of fire is *the* pivotal activity that represents the transition of human beings from nature to culture, and hence to civilisation, with the discovery of cooking being "the decisive factor in leading man from a primarily animal existence into one which is more fully human" (cited in Symons 2004, 104). Importantly, the discovery and development of cooking enabled human beings to absorb far more nutrients than is possible from a diet of raw food:

> The overarching benefit of cooking is that it acts as a kind of predigestion that extends the human body's ability to extract nutrients efficiently, greatly increasing our ability to adapt to changing circumstances...We use technology to process food according to the needs of the moment and store up what we learn in the form of tradition. *The resulting body of knowledge represents the most successful innovations, refined by experience and accumulated over generations.* (Gremillion 2011, 26; italics added)

V. A. Swinbank, *Women's Food Matters*,
https://doi.org/10.1007/978-3-030-70396-7_2

There exists a considerable body of literature on the historical development of the uniquely human activity of cooking (Wrangham 2009; Standage 2009; Fernandez-Armesto 2004). However, none acknowledge the historical role of women in this vital civilising activity, either crediting 'man the hunter' with the discovery of fire, or rendering women invisible with the use of generic terms, such as 'mankind' and 'humanity'. In this chapter, I 'flip-flop' mainstream thinking on this subject by addressing women's generally unacknowledged role in the creation and development of the world's myriad regional cuisines, and the rich culinary know-how and knowledge that has been the basis of human culture, civilisation and survival. Consequently, my feminist approach to the history of cooking is to invert the androcentric thinking that credits men with that most civilising activity—cooking—and to re/claim women's central role in the development and evolution of the world's food and culinary cultures.

An Alternative History of the Development of Cooking: Man the Hunter vs Woman the Gatherer

The development of cooking has been viewed through an androcentric lens in many historical accounts (Wrangham 2009; Standage 2009; Tiger and Fox 1972). What is less known is the central role that women have played in the role of food preparation and cooking throughout history. It is thought that the use of fire in the preparation of food was developed at least 400,000 years ago (Goudsblom 1992, 35). This development or discovery had a profound effect upon the development of the human species because it greatly increased the availability of foods, which could be rendered more digestible and often less toxic than in their raw state. A popular belief is that cooking was discovered by male hunters who found that meat accidentally cooked by a fire, started perhaps by lightening, tasted better and lasted longer than in the raw state, and from this invented the means to make this happen deliberately. However, it is now considered by a number of feminist scholars that the technology associated with food-gathering and cooking was developed by female hominids (Zihlman 1981; Ehrenberg 1989; Stanley 1981). The general belief that the early hominid's diet was a meat-based one from large animals provided by male hunters is now also known to be incorrect, and strongly disputed by mostly female historians and anthropologists, such as Evelyn Reed (1975), Margaret Ehrenberg (1989), Adrienne Zihlman (1981) and Autumn Stanley (1981). They conclude that women, in their role as mothers and food-gatherers were the principal providers of plant foods which traditionally make up the bulk of the diet in hunter-gatherer societies, and in that role invented or developed the first technology associated with preparing, processing and storing food, including the controlled use of fire. Feminist anthropologist, Adrienne Zhilman contends that gathering, not hunting was "the initial food-getting behaviour that distinguished ape from human" (Zhilman 1981, 93). She argues that hominid females developed the digging stick for grubbing up roots, a technology that enabled the later development of horticulture and agriculture. In order to free up their hands for food-gathering, they would have developed a sling to carry their infants whilst foraging, as well as containers in which to carry their foraged food. Zhilman points out that the !Kung hunter-gatherer women of the Kalahari still use this very technology, pointing to a similar technology practised by our pre-agricultural hunter-gatherer ancestors.

The bulk of the diet of these hunter-gatherers would have been plant-based, supplemented by protein from small animals and insects, and the occasional meat from hunting; food from plants was generally far more plentiful and reliable compared to the unpredictable nature of food supply from hunting. Again, the diet of the !Kung is made up mainly of plant foods and small animals, and as Zhilman points out, studies of hunter-gatherers, such as the !Kung "show the importance of women in these groups as an economic force, as they must have been in the past, though the evidence for this role would leave little or no trace in a record of the past" (93). Because the technologies—wooden digging sticks, food containers made of animal skins or woven plant material—were perishable, in contrast to tools made of stone, there are few, if any, archaeological records remaining of them. Conversely, because tools made from stone are associated with hunting large animals, it is generally assumed that it was male hunters who developed the earliest technology. However, even stone implements such as flints may have been used for sharpening sticks for grubbing out roots, as they still are amongst present-day hunter-gatherers. According to feminist researcher of early human technology, Autumn Stanley, women were responsible for developing the first technology for both cooking and agriculture:

> No one seriously disputes the overwhelming division-of-labour evidence that women invented cooking...(with) women's clever inventions of soap-stone griddles and pots, of waterproof cooking baskets and the use of hot stones to boil mush or liquids being the high technology of their day. (Stanley 1981, 292)

However, in the androcentric tradition of many academic texts on food preparation and cooking, men are credited with creating human culture, including the uniquely human activity of cooking. One notable exception to this is anthropologist Johan Goudsblom, who considers that cooking, which in the early stages of human development "was largely the work of women" and would have led to "social co-ordination and individual discipline, with useful spin-off effects as well", supplying the "first subtle and intimate knowledge of matter, thus forming the basis for the future development of the empirical sciences" (Goudsblom 1992, 35). According to Zhilman, the 'obsession' of most anthropologists with the image of 'man the hunter' has prevented them from looking at women's role in the development of human culture. She considers that this androcentric view credits 'man the hunter' with the development of tools, and for the socialisation of humans evolving from the supposedly co-operative behaviour needed for hunting in groups. In her view, a more plausible explanation for the development of human culture portrays:

> females as innovators who contributed more than males to the development of such allegedly human characteristics such as greater intelligence and flexibility. Women are said to have invented the use of tools to defend against predators while gathering and to have fashioned objects to serve in digging, carrying and food preparation. (Zhilman 1981, 98)

Despite these insights, the male-centred thinking that pervades much of the academic literature on the topic of food and cooking perceives women as passive recipients of the results of men acting on and shaping their environment (Wrangham 2009; Standage 2009; Tiger and Fox 1972). Such thinking equates women with nature and men with culture, a common misconception fuelled by stereotyped gender norms. German feminist sociologist, Maria Mies points out that the masculinist perception of female productivity is that it is unconscious and passive, especially women's ability to give birth; women are perceived as being closer to nature, and men as the creators and shapers of culture. This mindset aligns women's biological role in procreation and suckling of babies with the instinctual existence of animals, with women's domestic cooking seen to be simply an extension of this essentially animal-like feeding or nurturing role. Because of the deeply imbued thinking in male supremacist culture that equates men with culture, any activity when done by men, including that which is normally done by women, such as cooking, is automatically elevated to a higher level. As the famous anthropologist, Margaret Mead, so perceptively noted:

> In every known society, the male's need for achievement can be recognised. Men may cook, or weave, or dress dolls or hunt hummingbirds, but if such activities are appropriate occupations of men, then the whole society, men and women alike, votes them as important. When the same occupations are performed by women they are regarded as less important. (cited in Rosaldo and Lamphere 1974, xiii)

If women are indeed responsible for the 'invention' of cooking as the evidence suggests, then it could be argued that, far from being separate or one-step removed from culture, as prevailing masculinist thinking asserts, women are in fact central to and instrumental in the creation of culture. Cultural anthropologist, Sherry Ortner suggests that in their role in both the socialising of children—"the transforming of infants from mere organisms into 'cultured humans'—and the transforming of 'raw nature' into cooked food, women in fact could not be more representative of culture" (Ortner 1996, 80). Anthropologist, Johann Goudsblom further reinforces this point, stating that 'culture' is "now generally accepted in the social sciences as the technical term for referring to those aspects of behaviour that are 'learned, shared and transmitted'" (Goudsblom 1992, 4). According to this perspective, what women do in the daily round and routine of domestic cooking and in the socialisation of children undoubtedly qualifies as culture. However, it is the very everydayness of women's domestic cooking that causes it to be overlooked and not counted as culture. As culinary historian Michael Symons points out, the very "repetitiveness of cooking is part of the reason why many western intellectuals have snubbed it" (Symons 2004, 26). Downplaying the significance of domestic cookery in favour of 'grand' or '*haute cuisine*'—a world dominated by the professional cooking status of male chefs—is one of the primary ways that women's cooking work is devalued historically including in contemporary Western culture.

'Writers Have Hardly Noticed': Women's Invisibility in Writings About Food History

A number of books have been written in recent years dealing with the subject of cooking and its history (e.g. Standage 2009; Wrangham 2009; Fernandez-Armesto 2004). Interestingly, nearly all these books have been authored by men, despite the fact that, historically and cross-culturally, cooking has been and is still done principally by women. The exception to this is professional or '*haute cuisine*' which has traditionally been done by men, an issue I address in the following chapter. This follows a long-standing tradition of men theorising on a cultural activity, even when that activity is done by women. What is striking, however, about all these books is that despite, or indeed maybe because of cooking's central importance in human development and civilisation, women are either never or rarely mentioned as active agents or instigators in this crucially important development; it is 'humans' or 'mankind' that is credited, thereby obscuring women's central role. As women are invisible generally from history, so they are from most histories of cooking, despite their central role in cooking since time immemorial. Given the historical propensity of men to credit themselves with women's achievements and creativity, including that most female of activities, cooking, it stands to reason that much of the history of cooking has been wrongly attributed to men.

One notable exception to this is Australian culinary historian Michael Symons who notes that "…gastronomic thinkers, such as Jean-Anthelme Brillat-Savarin, have tended to pay more attention to enjoying than to preparing dinners" (Symons 2004, x). Symons is also a rare exception in his recognition of women's role in shaping the history of cooking, and hence civilisation:

> Cooks have been in charge –finding, sharing and giving food meaning. We could not have survived without them. They have been everywhere, yet writers have hardly noticed....Almost without exception, they have failed to enquire into the chief occupation of at least half the people who have ever lived…*Cooks have generally been women, and their achievements have been overlooked as inglorious and private…But while each of the cook's actions might be infinitesimal, the results have multiplied into civilisation.* (2004, x, italics added)

What is generally celebrated is the cooking of 'great' male chefs, not the daily domestic cooking of women, who throughout the ages have fed and kept humanity alive, making extraordinary meals out of ordinary and often limited ingredients. This has involved ingenuity, resourcefulness and creativity. The general belief/assertion that the peasantry of the world has subsisted on dull, monotonous gruel is an all-pervasive myth that ignores/denies/denigrates these female achievements.

Even when women write about the history of cooking they tend to concentrate on the cooking of men, for example Amy Trubek's "*Haute Cuisine*" (2000) and T. Sarah Peterson's "*The Cookbook That Changed the World*" (2006). A general exception to this, however, are female-authored cookbooks in which the rich legacy of women's home cooking is often acknowledged. As French cookery-book author, Madeleine Kamman states, acknowledging that women's home cooking has formed the basis of the world's myriad cuisines: "these women's dishes can be traced back almost to proto-history….The food lore put forth by women across the world is so rich a treasury that it can be tapped almost endlessly by the serious student of cookery in order to link the past to the present" (Kamman 2002, 6–7). In a similar vein, Greek cookery writer, Aglaia Kremezi, who dedicates her book to her mother and to the "Mediterranean women who bring love and joy to the table, the two secret ingredients that elevate even the humblest meal to an amazing feast", says that, when preparing the daily meal on the Greek island of Kea based on produce from her garden, she feels that every day she "can rely upon a tremendously rich legacy to guide me. The dishes I grew up with were created and perfected over centuries by resourceful female cooks from all over the Mediterranean, inventing myriad ways to use seasonally local produce" (2014, 12).

Despite the reluctance of most male food writers and historians to acknowledge this rich female cultural heritage, there is the occasional notable male cookbook writer who is willing to credit women's creative culinary genius, one such being the following, expressed by Ken Hom:

> Chinese cuisine reflects Chinese life itself: The grand design and profound richness of Chinese cuisine rests upon many centuries of dedicated application, splendid institutions, accidental discoveries, and *brilliant improvisations by millions of ordinary Chinese, usually women, working within countless family kitchens. Their skills and knowledge have been passed on through the centuries and spread gradually across villages and regions.* (Hom1990, 95; italics added)

Acknowledging the significance of women's culinary legacies is also a strong feature of Turkish writer, Tarhi Öztan's, *A Taste of Heaven in Turkey; the Gourmet Cuisine of Gaziantep,* based on five years of researching the cooking of thousands of women in and around Gaziantep in the south-east of Turkey. Speaking passionately about the cuisine of this region, Öztan sings the praises of the many generations of women who have created the vast number of dishes for which Gaziantep is famous. In his book recording this rich cultural heritage, he emphasises that "these dishes are the creation of Gaziantep women. They are the ones who made them famous" (Öztan 2014, 10).

Cooking and the 'Mating' Hypothesis

In contrast to the above, and more typical of male-authored histories of cooking is that by Richard Wrangham, whose book, '*Catching Fire: How Cooking Made Us Human*' (2009), is a classic example of male-centred, sexist theories on the evolution of cooking. Wrangham relies on gendered stereotypes about men and women which position women in subordinate roles. Amongst his many sweeping and unsubstantiated claims is his contention that cooking is the original basis of the sexual division of labour, and hence of women's subservience to men. He subscribes to the traditional masculinist image of 'man the hunter', who, on returning hungry from the hunt, expected to have a meal prepared for him by his female mate, who in turn needed his protection against the danger of other males stealing her food:

> A hunter-gatherer woman returns to camp in the middle of the day carrying the raw foods she has obtained. She then prepares and cooks for the evening meal at her own individual fire …*Females needed male protection, specifically because of cooking.* A male used his social power both to ensure that a female did not lose her food, and to guarantee his own meal by assigning the work of cooking to the female. (Wrangham 2009, 155–160; italics added)

Wrangham, who is a primatologist, says that his "interest in cooking comes largely from trying to understands reasons for the similarities and differences between the behaviour of chimpanzees and humans" (211). He draws parallels between non-human primates and humans with regard to food and mating, arguing that: "The food guarding, provisioning by females, and respect for possession found in animals are associated with males competing over sexual access to females, but only in humans have they led to households" (174). He even goes so far as to compare insect feeding behaviour with humans, based on observation of a particular Australian insect, that 'allows' a male to ride on her back in order to prevent other males from interfering with her own access to food! He concludes from this that "females feed males to reward them for behaving well. That is close to the system found in humans" (172). He asserts that cooking itself gave rise to the heterosexual male-female bond. Because cooking takes time, and the lone female cook needs to be able to defend against "determined thieves such as hungry males without their own food…Pair bonds solve the problem. Having a husband ensures that a woman's gathered foods will not be taken by others; having a wife ensures the man will have an evening meal" (155). I contend that Wrangham's hypothesis is highly problematic. Extrapolating the behaviour in non-human primates, and even insects, to humans is certainly questionable, to say the least. Furthermore, his theory of cooking perpetuates the androcentric, heteronormative myth of 'man the hunter' who, having been out hunting all day, returns to the hearth for a meal prepared by his female mate, a sexual dimorphism that "conjures up the image of big, strong, dominant men as hunters of large beasts and protectors of small, fragile women" (Zhilman 1981, 97). Wrangham's hypothesis assumes that an individual woman sat tending her individual fire as she cooked the evening meal for 'her man', imposing a modern individualistic notion on ancient human relations. Zhilman rightly warns against assuming "a division of labour such as occurs in living peoples to apply automatically to the ancient past" (105).

Wrangham's claims also call to mind the notion of individualistic rather than collective cooking. Rather than a lone woman attending her individual fire, it is far more likely women gathered food and cooked it co-operatively with other women, both for efficiency and possibly safety. The notion of the weak, defenceless female needing to be provided for and protected by a male partner is very much a product of Western masculinist thinking that tends to dominate most analysis of the evolution of human behaviour. According to Zhilman "contrary to many reconstructions of early human social life that picture women as burdened with young, sitting back at camp waiting for the hunter's return, hominid mothers must have been actively moving around the environment, getting food and carrying infants while doing so" (1981, 89). The survival success of early humans depended upon co-operation and sharing, rather than individualism and competition. Feminist anthropologists contend that sharing probably began between a mother and her offspring, with food sharing being mainly within the matrifocal group. As Zhilman further explains:

> Sharing requires physical proximity and would be encouraged by stability of group membership. The enduring nature of the primate mother –child tie suggests that groups of mothers and children would be the nucleus of permanent groups amongst hominids. Both sharing of food and sharing of infant and childcare are incompatible with rigid dominance hierarchies or extremely aggressive behaviour. If dominant individuals could usurp the food hunted or gathered by less dominant members of the group, there would be no real sharing. (9–10)

According to American anthropologist, Kristen Hawkes, the hunting hypothesis persists partly because "the association of stone tools with the bones of large animals at central places, dates to the beginning of the archaeological record, and coincides broadly with the origin of genus *Homo* itself" (Hawkes et al. 2002, 146). The prevailing belief that prehistoric male hunting provided the bulk of the early human diet is challenged by the fact that in existing hunter-gatherer societies, food from hunting is sporadic and unreliable, and that it is "women's foraging, not men's hunting, that differentially affects an individual family's nutritional welfare" (Hawkes et al. 2002, 247). According to Hawkes et al., the 'hunting hypothesis' that credits male hunters in prehistoric society with the evolution of human beings, including social learning, persists largely "because of the absence of an alternative". The authors contend that: "The grandmother hypothesis challenges the popular model in which men's hunting to provision wives and offspring is key to the evolution of distinctly human patterns of social organization and child development" (253). Hawkes et al. argue that the 'grandmother hypothesis' provides such an alternative: human beings are unique amongst primates in the fact that human females live past their reproductive years, so that post-menopausal women have long played a vital role in supporting their daughters during their child-bearing years with the feeding and socialisation of grandchildren .

'Crude and Indigestible': The Denigration of Women's Rural Cooking

Apart from the absence of women in most histories of food and cooking, there is a general assumption that the cooking of the common people consisted largely of barely palatable gruel, in contrast to the food of the wealthy, with a diet rich in often highly spiced animal protein. This assumption is based partly on the fact that there are few records of what the general populous ate, whereas there are many records of the diet of the rich and aristocracy. A typical view of the food of the common people is expressed, for instance, by food historian Clifford A. Wright who, although admitting that there is little knowledge of the diet of "the huge majority who lived on the margins of rich society", nevertheless claims that the poor "had no cuisine", subsisting on "rations of salt meat, hardtack, wine and vinegar" (Wright 1999, 48–49). He omits to say how poor peasant women managed to create amazing dishes despite the relative lack of costly ingredients, like meat. What little animal protein they had was supplemented by home-grown vegetables, wild plants, herbs and fungi. Italian ethnobotanist, Andrea Pieroni, for instance, found that women in Tuscany use more than fifty different wild plants to make soups, and amongst the Albanian Arbereshe people of southern Italy, Arbershe women use more than 110 plants, of which about fifty are wild species (Pieroni 2003). Nevertheless, despite this botanical richness and the resulting culinary and nutritional variety, Wright, like many others, assumes that the food of the poor was dull and monotonous; he considers the real impetus for the development of Mediterranean cuisines was the "creative impulse of the Renaissance" (41).

In a similar vein, in his book "*Delizia! The Epic History of the Italians and Their Food*" (2008), John Dickie refutes the general belief that Italy's cuisine is peasant-based. Describing the cooking of the 'masses' as "crude and indigestible" he asserts that, contrary to popular belief, Italian cuisine originated in the cities of Italy, not the countryside. Emphasising that "*Italian food is city food*", Dickie considers that:

> A history of Italian food written as the story of what peasants ate would make for a stodgy read. Many pages would be devoted to vegetable soup. There would be a substantial section on porridge. Bread made from inferior grains, and even from things like acorns in times of hardship, would need in-depth coverage.....It may have been the country folk who produced

> the cheese and pears, but the people with the power to appropriate these ingredients, and with the knowledge to transform them into a delicacy by a simple but artful combination, were the inhabitants of the *cities*. (Dickie 2008, 6–7, original italics)

Dickie's statement typifies a belief that only the food of the wealthy can be tasty and interesting. Whilst it is true that historically, and indeed even now, the 'pick of the crop' tends to be diverted from the country to the cities, and that urban centres have always had more access to different, more varied and imported ingredients, it is inaccurate to assume that means that the cooking of the countryside is necessarily inferior to that of the cities. The main difference is that country cooking is simpler and less rich in animal products. Nevertheless, the very simplicity for which Italian cuisine is renowned is attributed to the cities, not the rural population by Dickie who says that "Italian cuisine has a justified reputation for being delicious without being pretentious, complex or expensive… (and that) those values were first integrated into Italy's civilization of the table not by the peasants, but by the middle classes of the peninsula's many cities" (257).[1]

However, in contrast to Dickie's assertion, in her book, "*The Food of Italy*", food historian and cookbook writer, Claudia Roden (2014), describes how "Italian food is basically country cooking for large families, a combination of peasant food and the grand dishes of the nobility, that were (only) eaten by the peasants on special occasions – some only once a year at carnival time" (2014, 10). It is an over-simplification to say that Italian cuisine is solely a product of cities, as Dickie insists; as Roden points out, although particular dishes "may have a city stamp, …they have their roots in the land because town and country in Italy have always been closely bound" (10). Even if some of Italy's signature dishes were a product of urban life (and of course the cooking in the homes of the ordinary urban population would have been done by women), their origins would almost certainly have been rurally based. For instance, speaking of the Italian province of Piedmont, Roden describes how, despite there being a rich tradition of country cooking in the region, it was only the French-influenced cooking of the middle classes of Turin that was recognised, with the "extraordinary rich variety of country dishes (being) ignored until they arrived at the capital with the influx of peasants who were to form the new working class" (22). A classic example of a particular dish being associated with a city is the famous Genoese pesto;

although, as the name implies, it is associated with the capital of Liguria, Genoa, its origins undoubtedly are rural. In her book, "*Mediterranean Cooking*", food writer and cookbook author, Paula Wolfert traces the many regional variations of garlic and olive oil-based sauces around the Mediterranean. With archaeological records showing a peasant in antiquity pounding garlic, olive oil and herbs, over time this basic sauce manifested as *skordalia* in Greece with the incorporation of bread or potatoes, *tarator* in Turkey with the addition of walnuts or hazelnuts, *salmoriglio* in Sicily using olive oil, lemon juice, herbs, salt and garlic, *harissa* in Tunisia with the addition of hot pepper and cumin or caraway, *chermoula* in Morocco, using garlic, cumin paprika, parsley and fresh coriander, *aioli*, enriched with egg-yolks in Provence, and of course the famous basil, pinenut or walnut, olive oil and garlic sauce of Genoa, *pesto*.

The belief in the paucity and dullness of the diet of the general population—a belief promulgated, it has to be said, mainly by male food historians—is, unwittingly or not, a dismissal of the rich heritage of women's domestic cooking, on which in fact all the great regional cuisines of the world are based. In the passionate words of Madeleine Kamman:

> Known in America as 'ethnic' food, the foods of the humble people represent the enormously rich font of cuisine provided by all the women in the world. One look at the cuisine of all the nations will reveal that it exists everywhere: in the Orient, in the Islamic countries, through Africa, all over the Americas, and all over Europe. Humble earthy foods and food combinations were produced by generations and generations of women in an attempt to feed families economically with what was at hand. The French provincial and regional cuisines, sometimes called cuisine de terroir, cuisine de femmes, cuisine de misere, or in more gallant, more modern terms, cuisine du coeur, and the cucine casalinghe of all the regions and provinces of Italy belong to this category. ...*This pool of women's foods is our ancestral patrimony. It is the material we are all made of whether we are born rich or poor...* (Kamman 2002, 6; italics added).

The recording of recipes in the early days of literacy was done almost entirely by men, who were usually upper-class men of leisure and professional gastronomes. These men were eaters rather than cooks, and as such professed themselves expert in all things gastronomic. As a result, these male-authored records tend to not only reflect the diets of the wealthy, but also have the effect of disappearing or marginalising women's central role in the development of cuisines. As the Moroccan cook and author,

Fatema Hal, says: "Our cuisine, whether it is berber, arabo-muslim or jewish, is before all else a history of women, even if, curiously, it is men of the seventeenth century who wrote it" (Hal 2005, 10; my translation from the original French).

The masculinist mindset that cannot allow for women's central role in the development of the world's myriad cuisines even extends to crediting men with the creation of national cuisines. The disappearing of women from their central role in the development of cuisine, either deliberately or unconsciously, features throughout most of John Dickie's history of Italian cuisine, and is vividly illustrated in his discussion of Pelligrino Artusi's role in recording and codifying Italian cuisine. Despite acknowledging that many of the recipes were sent to Artusi by his literate middle-class female readers, thereby creating a 'multi-authored work', Dickie nevertheless asserts that "even today, Italians discover that their grandmothers' recipes are often, in reality, Artusian formulae that have seeped *unattributed* into family lore" (Dickie 2008, 196; italics added). I would argue, however, it is in fact, once again, women's domestic cooking that is being generally unattributed by male culinary historians, such as Dickie. Also, the point here is that, although Artusi was concerned with recording and codifying Italian cuisine, he did not *invent* the recipes he recorded. The development of cuisine is not the result of an individual 'invention', but the accumulated, shared knowledge of women's *collective* creativity over many generations in their daily role of domestic cooking.

In his description of the development of cooking during the Renaissance, Dickie credits Bartolomeo Scappi, the cook to the pope, with having "had a profound influence over Italian eating for centuries…and he left us with a picture of a cuisine that can justifiably be called *Italian* – a cuisine, in other words, that drew on ingredients and techniques from most corners of the peninsular" (102; original italics). He considers that: "Cooks are the heroes of the history of Italian food in the late Middle Ages and the Renaissance", with most of them coming from humble rural backgrounds, and therefore having a knowledge of the produce brought into the cities from the countryside. It goes without saying that these 'heroes', who became employed in the households of the aristocracy and the wealthy, were male. The fact that they came from humble rural backgrounds means that they would certainly have learnt their cooking and food knowledge and skills from observing the cooking of the local women, particularly their mothers and grandmothers, but this is not acknowledged by Dickie. He mentions the fact that hundreds of these (male) cooks who "brought Italian food to such heights" remain unnamed and unknown, but the truly unnamed and unacknowledged are in fact the countless generations of female domestic cooks who provided the culinary background to these professional male chefs.

The assumption that the diet of the peasantry consisted of little more than monotonous tasteless gruel is a common theme in most—especially male-authored—histories of food. In fact, whilst the cooking of the wealthy aristocracy may have become codified and hence standardised, regional peasant cooking, due to its very frugality, was often highly innovative. Indeed, as food historian and cookery-book writer, Elisabeth Luard asserts, the very limitations imposed on peasants, with their dependency on what they grew or was locally available gave it its "particular identity ... and in no sense meant that the ingredients were necessarily poor or inferior" (Luard 1987, xiv). Speaking of the harsh terrain and poverty of the Riviera, especially the inland region, food writer Colman Andrews contends that the "very poverty of the region's cuisine mandated a certain rustic inventiveness, a culinary creativity grown ...out of the sheer need to survive (and that) the indigenous traditional cuisine of the Riviera – as opposed to hotel and fancy restaurant fare and more contemporary refinements – is based on ingenuity, conservation, and reuse (of leftovers)" (Andrews 2000, 12). He describes two wonderful meals, both prepared by women, one on the French Riviera and the other on the Italian Riviera, based only on what was locally available; the Italian one, for example, consisted of several simply prepared dishes of local vegetables and herbs and walnuts, and a rabbit stew with herbs and white wine—in short, the local, humble ingredients that local rural women would have been cooking for generations. He comments that he was "astonished and impressed. Everything seemed so honest and direct – so much of the region, and confident in its regional identity. And everything tasted wonderful. This was real food, and not food that I was likely to encounter anywhere else in the world" (3).

'Inspired Improvisation': The Rich Tradition of Female Domestic Cooking

In '*Puglia: a Culinary Memoir*', Maria Pignatelli Ferrante describes how the cooking of the region is "the fruit of thousands of years of poverty, and the creativity and intelligence of a people struggling simply to survive....creating simple and inventive recipes, which call on all the cultivated and wild produce that grows in this ancient land" (Ferrante 2008, 7–10). She describes how the women of Puglia have shown enormous creativity in inventing dishes that are "nothing short of true gastronomic masterpieces", such as a vast variety of foods fried in the plentiful olive oil of the region, and 'tielle', casseroles of layered baked vegetables (53). Pulses, such as fava beans, both fresh and dried, chickpeas, lupini beans and lentils provided most of the protein of the peasantry; the potential monotony of such staples was enlivened by the imaginative addition of vegetables, such as chicory, and a wealth of other greens and herbs, both wild and cultivated, fungi gathered in the wild, and chilli once it was introduced from the New World. Ferrante describes how the household vegetable garden provided zucchini, artichokes, peppers, tomatoes and eggplants, with the surplus being preserved in olive oil, or dried in the sun, to be re-hydrated in winter. Preserved olives and anchovies also added variety and flavour. Meat was a relative luxury, eaten only occasionally, such as on festive days, although wild rabbit and a variety of preserved meats, 'salume', from the family pig, as well as local cheese, such as sheep's milk pecorino, provided additional protein, and the flavours so characteristic of the region. Far from being the dreary, monotonous diet of the peasantry described by food historians, such as Dickie, the food of one of the poorest parts of Italy, Puglia, illustrates how the ingenuity and imagination of the women cooks of that region created a delicious, distinctive, healthy and richly varied cuisine based only what was locally available. As Ferrante points out, this peasant food, or '*cucina povera*' has now become all the rage in fashionable restaurants.

Calabria, another region of southern Italy known for its often harsh and challenging terrain, and its traditional '*cucina povera*', is rich in biological diversity "embodied in a cultural diversity closely bound up with the methods adopted for processing the local produce" (Olivera 2014, 15). Its cuisine is a product of people wresting a living out of its challenging terrain, using creativity and inventiveness, "devising new ways and techniques for extracting flavour from simple ingredients" (15). Far from the local cuisine being dull and monotonous, Valentina Olivera observes that "it is impossible to find a single recipe that unifies the region. There exist countless variations and local versions of every dish, growing out of the complex interweaving of history and nature" (15). In Crete, also a traditionally poor part of the Mediterranean, meat has constituted a small part of the traditional Cretan diet, considered to be one of the world's healthiest. The diet of the Cretan peasantry was based on grains—wheat or barley—pulses, vegetables and olive oil. In addition, for religious reasons, meat was forbidden for at least one-third of the year, although fish was allowed. As a result, the cooking relied on vegetables, herbs, wild plants, grains and pulses, which, far from making for a limited and boring diet, inspired innovation in creating an enormous variety from seasonally available produce. In Cretan cooking "All the flavours come together harmoniously in a dish and show a fine balance of taste.... Cretan food is a way of life, simple and basic...and quite delicious! *Combined in this simplicity is a creativeness of the housewife who has in her hands centuries of experience.... The women of Crete are the ones to whom we now owe the survival of ancient recipes along with their creativity*" (Psilakis and Psilakis 2000, 24, italics added).

The common assumption that the harsher the environment, the poorer the diet is, is refuted by French food historian, Jean-Louis Flandrin. He explains that, in a fascinating paradox, peasants in more geographically challenged places were often healthier and also more resilient against food shortages. This was because people in fertile areas tended to focus on grain growing, such as wheat, which constituted the bulk of the diet, with the surplus being sold on the market. As a result, in times of crop failure they were much more likely to suffer from famine than those living in less fertile environments. He gives a couple of examples, both French. The inhabitants of the Beauce region, one of Frances's richest breadbaskets, with their reliance on wheat, were vulnerable to crop failures, during which they were forced to seek refuge with the peasants of the neighbouring, geographically poor Solonge region who fared better due to their "comparatively old-fashioned and therefore more varied diet". Similarly, the peasants of the wealthy and fertile Limagne region had generally poor health as a result of their reliance on wheat and suffered during crop failures. In contrast, the peasantry of the Auvergne with its poor soil and harsh climate were much healthier because of their much more varied diet based on raising livestock, hunting, gathering wild plants and fishing, as well as farming (Flandrin 1999, 352).

Far from stunting culinary imagination, the very nature of the relative limitation of ingredients created by using what was only locally and seasonally available was, in fact, an impetus to creativity and innovation. As Tom Stobard reminds us: "We should remember, too, that inspired improvisation by cooks in circumstances where only a limited range of ingredients was obtainable led to the invention of some of the world's great dishes" (1980, intro.). Although he does not mention the sex of the cooks, they would inevitably have been female as professional male cooks generally have not had to deal with scarcity. Greek food writer, Aglaia Kremezi, points out that: "Frugal cooks around the Mediterranean were limited to a small variety of vegetables and greens for months on end, so they devised an incredible number of ways to prepare them differently. They stew vegetables with aromatics, stuff them with bulgur or rice, grate them, and mix them with cheese to make the filling for pies, or fry them and serve them with tarator or skordalia" (Kremezi 2014, 128). Yet another example of women's ingenuity, in making sure that no precious food is wasted, is the vast repertoire of cooking based on the use of leftover stale bread. Describing how "an entire cuisine has been built on leftover bread", Carol Field lists the enormous range of dishes made from stale bread, including soups, salads, dumplings, desserts, with a myriad variation depending on the region and the individual cook (Field 1997, 18). Describing the food culture of the Mediterranean region, food historian Carol Helstosky, notes that under conditions of food scarcity caused by poverty, and during the aftermath of the Second World War, women have had to be particularly inventive, for instance, making 'meatballs' out of stale bread. She considers that: "Under such conditions, cooking is a creative achievement that is frequently overlooked or undervalued by historians of food and cooking, who tend to focus on the more elaborated culinary achievements of individuals or societies" (Helstosky 2009, 65).

Italian food writer Anna del Conte explains that the appreciation and respect Italians have for the quality of ingredients, with the main flavour of the produce "coming through loud and clear", is directly connected to the love Italians have for '*cucina casalingua regionale*' (regional home cooking), and that "even recipes developed by the greatest chefs can be traced back to home cooking" (del Conte 2004, 10). Indeed, the fact that the best and most authentic regional cooking is invariably found in the homes of the common people is generally a cross-cultural phenomenon. Speaking of Indian cooking, for instance, Priti Narain contends that "of course the best food from any region is to be found in the homes of the people. Each family has its own way of cooking a particular dish and each one is 'authentic'" (Narain 2000, 4). Similarly, albeit in a different cultural context, Turkish food historian, Tekin Öztan, describes how, in his extensive research on the domestic cooking of Gaziantep in southeast Turkey, he found that, faced with thousands of recipes, "we sorted them by dish, only to find out there were a hundred different versions of each....Two neighbours, even two siblings could give you different recipes for the same dish" (Öztan 2014, 11).

Anthropologist Jack Goody points out that the emphasis on high meat consumption has resulted in the belief by many food historians that meat is the most important element of the human diet. An accompanying assumption is that the diet of the peasantry, with its traditionally low meat consumption is poor and inadequate. In fact, in general, the diet of the peasantry was probably a lot healthier than that of the wealthy. As Elisabeth Luard says: "The old peasant kitchen habits of frugality (involved) making stock out of bones, pickling and salting in times of glut, stocking the pantry, (and) making good food out of few and simple ingredients" (Luard 1987, xiv). She adds that "in the real terms in the life expectancy of the peasantry, such as that of England in feudal times, who, having survived the dangerous childhood years, were likely to live longer and in better health than their overlord who dined daily on large quantities of meat and white bread" (xiv). This, of course, is not to deny the fact that there were not times of great scarcity and hardship for the peasantry, be it from poor weather, or more likely, the fact that landlords purloined the bulk of the peasants produce in many regions. Peasant, that is women's cooking, used meat—admittedly per force—in small quantities. In the diet of the wealthy where meat was eaten in large quantities, with a preference for roasting, it was a symbol of status and power (Montanari 1999, 180). Whilst meat was highly esteemed, vegetables, and especially legumes, were generally regarded as the food of the poor and so were not valued. Vegetables generally came from kitchen gardens; eggs and honey were also home-produced and therefore were outside the money economy (Santich 1995, 24). Significantly, because women were the producers of these home products they were not seen to have market value, despite rural women selling any surplus in the village market. This devaluing or lack of acknowledgement of women's major contribution to the development of the world's cuisines is a manifestation of the masculinist devaluing of women's creativity and ingenuity, whilst attributing these qualities exclusively to men.

Women's frugality and ability to make do with whatever is available tended to be dismissed, even disparaged by male chefs. Renowned French chef, Auguste Escoffier, for instance, opined that a "woman...will manage with what she has handy. This is very nice and obliging of her, no doubt, but it eventually spoils her cooking, and the dish is not a success....her cooking pales before that of a man, who makes his dishes preferable on all occasions to hers" (cited in Trubek 2000, 125). In the following chapter, I examine the development of a culinary hierarchy in which male professional cooking is awarded a higher cultural status than the female tradition of domestic cooking, even though the former is dependent, although generally unacknowledged, on the latter.

Note

1. Although cities may not be the main source of regional cuisines, as Dickie insists is the case with Italian cuisine, it is important to note that, historically, many cities in the world, especially those on trading routes, have often provided a rich melting pot of different cuisines where women from different ethnic and religious backgrounds often exchange recipes. One such example is that in Nawal Nasrallah's book on Iraqi cooking. She describes how, when she was growing up in Baghdad, such sharing and exchange took place in her neighbourhood between women including those who were Kurdish, Armenian, Palestinian, Indian, Iraqi of course, as well as who were Muslim, Jewish and Christian (Nawal Nasrallah, *Delights from the Garden of Eden: A Cookbook and History of the Iraqi Cuisine*, 2013, Equinox Publishing, Sheffield, UK).

References

Andrews, Colman. 2000. *Flavours of the Riviera: Discovering the Real Cooking of France and Italy*. London: Grub Street.

Del Conte, Anna. 2004. *Classic Food of Northern Italy*. London: Pavillion.

Dickie, John. 2008. *Delizia: The Epic History of the Italians and Their Food*. New York and London: Free Press.

Ehrenberg, Margaret. 1989. *Women in Prehistory*. London: British Museum Publications.

Field, Carol. 1997. *In Nonna's Kitchen: Recipes and Traditions from Italy's Grandmothers*. New York: HarperCollins.

Flandrin, Jean-Louis. 1999. "Introduction: The Early Modern Period." In *Food: A Culinary History*, edited by Jean-Louis Flandrin and Massimo Montanari, 349–373. New York and London: Penguin.

Fernandez-Armesto, Felipe. 2004. *Near a Thousand Tables: A History of Food*. New York: Free Press.

Ferrante, Maria Pignatelli. 2008. *Puglia: A Culinary Memoir* New York: Oronzo Publications.

Goody, Jack. 1982. *Cooking, Cuisine and Class: A Study in Comparative Sociology*. Cambridge: Cambridge University Press.
Goudsblom, Johan. 1992. *Fire and Civilisation*. New York: Penguin.
Gremillion, Kristen J. 2011. *Ancestral Appetites: Food in Prehistory*. Cambridge: Cambridge University Press.
Hal, Fatema. 2005. *Le Grand Livre de la Cuisine Marocaine*. Paris: Hachette.
Hawkes, K., J. F. O'Connell, N. G. Blurton Jones, H. Avarez, and E. L. Charnov. 2002 "The Grandmother Hypothesis and Human Evolution." In *Adaptation and Human Behavior: An Anthropological Perspective*, edited by L. Cronk, N. Chagnon, and W. Irons, 231–252. New York: Aldine de Gruyter.
Helstosky, Carol. 2009. *Food Culture in the Mediterranean*. Westport, CT and London: Greenwood Press.
Hom, Ken. 1990. *The Taste of China*. New York: Simon and Schuster.
Kamman, Madeleine. 2002. *When French Women Cook: A Gastronomic Memoir*. Berkley, CA: Ten speed Press.
Kremezi, Aglaia. 2014. *Mediterranean Vegetarian Feasts*. New York: Stewart, Tabori and Chang.
Luard, Elisabeth. 1987. *The Old World Kitchen: The Rich Tradition of European Peasant Cooking*. New York: Bantam Books.
Mies, Maria, Veronica Beenholdt-Thomsen, and Claudia von Werlhof. 1988. *Woman, the Last Colony*. London: Zed Books.
Montanari, Massimo. 1999. "Peasants, Warriors, Priests: Images of Society and Styles of Diet." In *Food, A Culinary History*, edited by Jean-Louis Flandrin and Massimo Monantari, 178–185. New York: Columbia University Press.
Narain, Priti. 2000. *The Essential Delhi Cookbook*. New Delhi: Penguin.
Olivera, Valentina. 2014. *Calabria in Cucina: The Flavours of Calabria*. San Vendemiano: Sime Books.
Ortner, Sherry. 1996. "So, is Female to Male as Nature is to Culture?" In *Making Gender: The Politics and Erotics of Culture*. Boston: Beacon Press.
Öztan, Tahir Tekin. 2014. *A Taste of Heaven in Turkey: Gourmet Cuisine of Gaziantep*. Istanbul: Oglak.
Peterson, Sarah T. 2006. *The Cookbook that Changed the World: The Origins of Modern Cuisine*. Stroud, UK: Tempus Publishing.
Pieroni, Andrea. 2003. "Wild Food Plants and Arberesh Women in Lucania, Southern Italy." In *Women and Plants: Gender Relations in Biodiversity Management and Conservation*, edited by Patricia L. Howard. London and New York: Zed books.
Psilakis, Maria, and Nikos Psilakis. 2000. *Cretan Cooking*. Heraklion, Crete: Karmanon.
Reed, Evelyn. 1975. *Woman's Evolution from Matriarchal Clan to Patriarchal Family*. New York: Pathfinder Press.
Roden, Claudia. 2014. *The Food of Italy, Region by Region*. London: Square Peg.

Rosaldo, Michelle Zimbalist and Lamphere, Louise, eds. 1974. *Women, Culture and Society*. California: Stanford University Press.

Santich, Barbara. 1995. *The Original Mediterranean Diet*. Adelaide: Wakefield Press.

Standage, Tom. 2009. *An Edible History of Humanity*. New York: Walker Publishing.

Stanley, Autumn. 1981. "Daughters of Isis, Daughters of Demeter: When Women Sowed and Reaped." *Women's Studies International Quarterly* 4 (3): 289–304.

Stobard, Tom. 1982. *The Cook's Encyclopedia*. London: Papermac.

Symons, Michael. 2004. *A History of Cooks and Cooking*. Chicago: University of Illinois Press.

Tiger, Lionel, and Robin Fox. 1972. *The Imperial Animal*. New York: Dell.

Trubek, Amy. 2000. *Haute Cuisine: How the French Invented the Culinary Profession*. Philadelphia: University of Pennsylvania Press.

Wolfert, Paula. 1996. *Mediterranean Cooking*. London: HarperCollins.

Wrangham, Richard. 2009. *Catching Fire: How Cooking Made Us Human*. London: Profile Books.

Wright, Clifford A. 1999. *A Mediterranean Feast: The Story of the Birth of the Celebrated Cuisines of the Mediterranean*. New York: William Morrow.

Zhilman, Adrienne, L. 1981. "Women as Shapers of Human Adaptation." In *Woman the Gatherer*, edited by Frances Dahlberg. New Haven and London: Yale University Press.

CHAPTER 3

Culinary Hierarchy: From Peasant Cooking to '*Haute Cuisine*'

The division between women's domestic cooking and professional cooking done by men, what I refer to as culinary hierarchy, with the latter generally given a higher status than the former, has a history going back several thousand years (Swinbank 2002). According to cultural anthropologist, Jack Goody, a sexual division of labour with regard to cooking has existed since the early Egyptian period, when "historical evidence for the emergence of a high and low cuisine first appears" (Goody 1982, 193). A culinary hierarchy is the product of a socially stratified society in which the wealthy seek to distinguish their diet and way of eating from the poor or common people. An additional feature is the association of '*haute cuisine*' with professional male cooks and the cooking of the common people with female cooks. Goody notes that the "difference between high and low cuisine tended to be one between male and female, in which men transformed the female recipes of daily cooking into the '*haute cuisine*' of the court" (193), a situation which has continued throughout history in the great courts of Europe and the Mediterranean. There are endless manifestations throughout the world of the creativity and ingenuity demonstrated by women's home cooking. This is what has formed the basis of all cuisines both 'low' and 'high', although the latter has been performed by men appropriating and elaborating on the former, that is, women's domestic cooking, although this is rarely, if ever, acknowledged. The main feature of what I refer to as a 'culinary hierarchy'

V. A. Swinbank, *Women's Food Matters*,
https://doi.org/10.1007/978-3-030-70396-7_3

is that 'high' or '*haute cuisine*' refers to professional cooking, traditionally done by men, and 'low' cuisine refers to domestic cooking done by women.

The appropriation and elaboration of women's domestic cooking by professional male cooks, which has existed since at least the time of the pharaohs, forms the basis of what is generally known as '*haute cuisine*' or 'grande cuisine'. Nowhere is this truer than in the development of French '*haute cuisine*' which developed during the seventeenth and eighteenth centuries in the kitchens of the aristocracy. This was not the first expression of 'grande cuisine'; it has existed in many other hierarchical societies ruled by an aristocracy and royal court—examples include Ottoman, Moghul, Persian and Chinese. A central feature of such cuisine is elaboration, using rare and expensive, often imported ingredients, such as spices, and an emphasis on high consumption of animal products. Such conspicuous consumption was a deliberate attempt to emphasise a social hierarchy separating the wealthy elite from the general populace.

In an interesting and important essay on the relationship between 'high' and 'low' cuisine, anthropologist Sidney Mintz describes "*haute cuisine*, so called, (as) some sort of refinement of the aggregate foods, styles and dishes of a collection of regions, a skimming off of representative foods (from different regions)" (Mintz 1996, 104). Furthermore, there "is a self-consciousness connected with an *haute cuisine*, and some sense that it transcends local difference" (101). One striking example of this process is that of Ottoman cuisine. As Sarah Woodward points out in "*The Ottoman Kitchen*", the famed cuisine of the Ottoman court which was dedicated to the pleasures of the table, was not invented by the chefs of the palace kitchens, but was largely a result of a 'plundering' and elaboration of regional dishes from the far reaches of the empire: baklava from Armenia, egg and lemon sauces from Greece, pastries and boreks from the Balkans and Central Asia.[1] A striking example of this is the famous Turkish dish Circaissian chicken, a sumptuous chicken salad with a walnut sauce, thought to have been brought to the Topkapi Palace by Georgian women procured for the sultan's harem. This is also affirmed by German professor of Iranian Studies, Bert Fragner, who says that "there is no doubt that culinary trends and fashions in the Ottoman Empire were very much shaped by Constantinople's metropolitan elites, often based on regional or local cuisines from various parts of the empire, but homogeneously shaped by the prestigious and refined taste of fashionable dandies in the vicinity of the Saray" (Fragner 2000, 52). Interestingly,

Turkish food writer and historian Ayfer Unsal says she is very irritated when people speak of Ottoman cuisine. She asserts that there is, in fact, no such thing, pointing out that no recipes from the Topkapi Palace kitchens actually exist; ingredients were recorded, but no recipes (Unsal 2015).

Cuisine as an Expression of Class

According to Sidney Mintz: "*haute cuisine* need not have geographical roots: its social character is based on class" (Mintz 1996, 101). Not having geographical roots in one region, but being made up of an amalgam of recipes from different regions, 'high' cuisine acquires a national status and sometimes even, like French '*haute cuisine*', an international status. This is in contrast to regional cuisines, which are products of local factors such as soil and climate, what the French call *cuisine du terroir.* Such regional or 'low' cuisines are, according to Mintz, more authentic than high or 'haute' cuisines, which, he considers, are, "like it or not, 'restaurant food' of the sort that turns up in restaurants abroad, and in capital cities" (104). This observation helps explain some of the differences between French and Italian restaurant cooking: with the development of '*haute cuisine*' in France, French cooking became regarded as an elite masculine profession, in contrast to Italy where small family-run restaurants have generally depended upon regional cooking done by women (Harris and Giuffre 2015, 32).

A striking example of a recent male chef's appropriation of women's culinary knowledge and expertise is that of the 'discovery' by the renowned Chilean chef, Guilleromo Rodriquez, of *merquen*, a traditional indigenous plant of the Mapuche. As ethnobotanist Patricia Howard (2003) explains, the international 'top chefs' society, Les Toques Blanches, has collaborated with Mapuche women, observing the women gathering *merquen* and other plants in the forest, and cooking traditional recipes using them. Rodriquez and fellow Chilean celebrity chef Rodolfo Guzman have incorporated *merquen* into their gourmet cuisine, claiming to have 'discovered' and 'rescued' it from indigenous cuisine. According to Chilean social anthropologist Mihalis Mentinis (2016) "in order to be incorporated within *haute cuisine*, *merquen* needs to be 'surrounded', 'combined', 'assembled' and presented with other elements", becoming 'sophisticated' whilst still retaining a 'strange rusticity' (56). As a result, it has been elevated from "the 'low' cuisine of the poor to the *haute cuisine* of the presidential palace and the restaurants of the elite" (55). Consequently, despite the fact that "it maintains a certain relation to indigenous culture, it no longer belongs to any particular group" (57). In 2010 UNESCO inscribed traditional Mexican cuisine as a 'cultural heritage of humanity', based on its "ancestral, ongoing community culture with emphasis on local, native, traditional, and feminine aspects of food culture" (Hryciuk 2019, 5). However, this recognition was promoted by state and tourist agencies as 'new Mexican gastronomy', presented as "cosmopolitan, transnational, innovative, and predominantly masculine....(which) once again, gentrifies peasant cooking served to an (international) elite becoming an accepted form of national heritage" (Hrycuik 2019, 5). The Mexican state of Oaxaca was particularly promoted in recognition of its rich culinary cultural heritage, with some *cocineras traditionales* (traditional female cooks) selected to work with famous male chefs helping to prepare meals at high-end restaurants during festivals. Nevertheless, although:

> intended as an opportunity to exchange knowledge and work together, the arrangement resulted in the chefs learning new cooking techniques, ingredients and local recipes, with only very few female cooks benefitting by gaining connections to the world of high-end gastronomy. The general impression was that, once again, the *cocineras* were treated as a highly skilled, ethnic workforce. The festival poster perfectly illustrated this inequality: photographs of unnamed indigenous women provide a backdrop for the images of famous male chefs. (Hrycuik 2019, 17)

Furthermore, the promotion of 'star' male celebrity chefs, both local and from abroad, presented an individualistic approach to cooking as the creators of the 'new' Oaxacan cuisine (*cocina de autor*). Such individualism stands in stark contrast to regional culinary traditions, such as that of Oaxaca, which are the product of many generations of women's *collective* knowledge and expertise.

French historian, Jean-Francois Revel also acknowledges that 'female collective low', or what he calls 'popular' cuisine forms the basis of 'high' or *haute cuisine*: "It suffices to leaf through the great treatises on cuisine from the last hundred and fifty years to note that they have adapted the principal ideas, but not necessarily the exact recipes of various regional cuisines" (Revel 1982, 223). He considers that whilst 'popular cuisine' "has the advantage of being linked to the soil...and in close accord with nature and based on age-old skills", these are "transmitted unconsciously by way of habit, by way of imitation and habit...prescribed by a long tradition" (19). However, he considers that what he calls 'erudite cuisine' is 'conscious' and "based, by contrast, on invention, renewal and experimentation". Furthermore, only gastronomy or '*haute cuisine*' "belongs to the domain of art, it alone is capable of comprehending the *creative principle* behind this or that local knack of preparing a dish and of applying consciously what was executed unconsciously or mechanically" (216, original italics). This implies of course, that family cooking, that is women's cooking, is unimaginative and 'unconscious'—in other words, unthinking—and thus inferior. Professional, that is men's cooking, is, by contrast 'conscious', innovative and creative. The belief of chauvinists, such as Revel, that women's domestic cooking is not imaginative or innovative, and "clings to its errors as to its qualities" is contradicted by the statements of those, usually women, who know first-hand just how creative and inventive women's cooking is. Unlike male chefs who are always anxious to distinguish their cooking as 'original' by tweaking traditional dishes and 'inventing' new ones, women's domestic cooking is a collaborative one, based on recipes handed down through the female line, incorporating new ideas and ingredients over time and by osmosis. Greek cookery author, Aglaia Kremezi explains that, far from being static, peasant cooking in her native Greece is a: "deep, multilayered story" which, although slow to change, incorporates new ingredients, such as tomatoes and peppers from America. As she says, peasant cooking "does not know the word 'fusion', but it does evolve. The strong attachment to tradition, with small openings for change and incorporation of new ingredients and techniques is the lifeblood of our cuisine. *Siga, siga*, we say often in Greece: 'Slowly, slowly'. But, in time, we embrace products and techniques from faraway places. The dialogue between old and new shows how Mediterranean cooks reflect deeply upon, and constantly improve, their traditional plant-based diet" (Kremezi 2014, 10).

Nevertheless, adoption of new ingredients by peasants is not always a slow process; Claudia Roden, in her encyclopaedic "*The Food of Spain*" describes how the peasants of the Iberian peninsula enthusiastically embraced the use of peppers when they were brought back from the New World. Similarly, in many parts of Africa, especially West Africa and maritime regions, both chillies and capsicums were rapidly adopted from the Americas, with the plants being incorporated in women's homegardens and integrated into the local cuisines (McCann 2010, 60). Regional cuisines are not frozen in time, as Revel insists, but are influenced by many external factors, including trade and conquest, but, perhaps most importantly, by movement of people due to migration, whether by choice or necessity. One such example is the enforced movement (after the fall of the Ottoman Empire) of ethnic Greeks from Smyrna in Turkey to Greece who brought with them Turkish flavours and spices they had absorbed into their cooking. There are countless other examples of this cross-cultural culinary influence throughout the world: for instance, the development of Nyonya cuisine (also known as Straits Chinese cuisine), a combination of Chinese and Malay cooking, resulting from inter-marriage between immigrant Chinese and ethnic Malays. More examples include the cooking of Oman, with influences from Africa, the Middle East and India as a result of being a focal point for trade from all these places.

'National Cuisines' vs Regional Cuisine

Sidney Mintz asserts that: "A 'national cuisine' is a contradiction in terms: there can be regional cuisines but not national cuisines" (Mintz 1996, 104). Confirming Mintz's definition of cuisine as being a product of a particular region's climate, ecology and history, rather than an artificial national entity, food writers Jeffrey Alford and Naomi Duguid comment that: "A map of the world would look very different, if, instead of marking political boundaries, we outlined regions based on food and culture" (2008, 1). According to this definition, there is no such thing as, for instance Chinese cuisine, but many different regional cuisines; apart from the famed regional cuisines of the dominant Han regions of eastern China—Cantonese, Peking, Sichuan—there are many other distinct regional cuisines from the many different ethnic groups that lie within the Chinese national border, including Dai, Dong, Hani, Hui, Kazakh, Kirgiz, Mongol, Tajik, Tibetan, Miao/Hmong, Tibetan, Uighur, Yi and Tuvan. The same applies to many other so-called 'national cuisines'. Italian cuisine, for instance, is in fact made up of twenty distinct regional cuisines stretching from Lombardy, Piedmont, Trentino-Alto Adige and Valle D'Aosta in the north, to Puglia, Calabria and Sicily in the south, all influenced by their particular climate and geography as well as cultural, including linguistic, and historical differences. Similarly, Indian cuisine consists of numerable regional cuisines, all very much influenced by climatic, geographic and cultural differences: Punjabi and Kashmiri in the north, Kerala and Tamil Nadu in the south, Gujarat and Goa in the west, Bengal in the east.

A central feature in the development of high or *haute cuisine* was the existence of literacy (Goody 1982). Whilst the replication of recipes in written form made them available for testing and improvement within a community of professional male chefs, it also had the potentially opposite, stultifying effect of diminishing improvisation, resulting in conformity and conservatism (Mennell 1985). On the other hand, the oral tradition, which characterises women's traditional cooking of passing on recipes and food knowledge and skills (which I discuss in Chapter 5) is more likely to lead to personal interpretation and innovation, the very opposite of the supposed standardisation critiqued by Revel. Talking about the regional cooking of India, for example, Monisha Bharadwaj (2006) explains that regional dishes have many variations because "each household will often have their own recipe for the same dish, one that has been passed down over the generations in that home. Over time, it will have been adapted so that hundreds of recipes exist for the same dish" (6). And, as Turkish ethnobotanist, Fusun Ertug explains: "'Traditional village food' is not constituted simply by the repetition of ancient recipes. Each woman has her own preferences and specialities as well as her collective or personal memory" (Ertug 2003, 188). In a similar vein, in a personal conversation with an Iranian woman, she remarked that in Iran every household cook has her own interpretation or version of traditional Persian recipes of which they are very proud. Indeed, Persian cuisine, noted for its exquisite refinement, is, again, the creation of women, with "thousands of recipes having been devised by Persian women" (Hekmat 2005, 8). The great historian of Mediterranean, Middle Eastern and Jewish food cultures, Claudia Roden (2014) asserts that "there is no one way with traditional home cooking and no strict rules...cooking does not stand still, it evolves while retaining its character. In Italy every family has their own way of making a dish, but it is always within certain norms" (8).

Interestingly, traditional female cooks, even when they could read, often distrusted written recipes, feeling that the "formalisation did away with the personal element" (Goody 1982, 148). Jack Goody gives the example of one such female cook, Magaridou from the Auvergne who, despite producing a 'special cookbook' herself, nevertheless "had a horror of cookbooks, (fearing) that the recipes of her grandmother might '*abimer en les ecrivant*" (get spoiled in the writing of them) (148). Similarly, Carol Field reports from her interviews with Italian grandmothers that they considered writing recipes as 'absurd', saying that: "There aren't recipes, there are ingredients... These are women who are confident in their food knowledge and in control at the stove: It is a rare grandmother who cooks from a written recipe. These women know what they are doing. They measure *all'occhio* (by eye), they scoop up *un pugno* (a handful, literally a fistful), and they pour *al dito* (by the finger)" (Field 1997, 26).

Michael Symons points out that throughout history "cooks generally have been women, and their achievements overlooked as inglorious and private...(they) have always been in the background – both ever-present and unnoticed". (Symons 2004, 25) However, despite being relegated to the background, women's culinary creativity and inspiration have been drawn upon, and appropriated by professional male chefs. At the same time, they have sought to distance themselves from women's domestic cooking, whilst denying their appropriation of and dependence on it. This illustrates what feminist philosopher, Val Plumwood (1993) calls 'background denial'; she defines this as "making use of the other, organising, relying on and benefitting from the other's services (while) denying the dependency this creates." This is done by "denying the importance of the other's contribution....and on insisting on a strong hierarchy of activities, so that the denied areas are simply not 'worth' noting" (48). Another feminist philosopher, Marilyn Frye, theorises that in phallocratic thinking, in order to maintain a sense of superiority, the 'master's' perspective requires both denial of his dependence on women, and denial and negation of women's reality (Frye 1983). This is precisely the dynamic underlying the existence of a culinary hierarchy. Women's domestic cooking is ignored or trivialised; yet, at the same time, '*haute cuisine*' depends upon it for its very existence, by drawing upon and appropriating this area of female creativity for its own ends. To ensure a separation and a denial of this dependency, a rigid hierarchy is created, which disowns any connection or debt to women's creativity.

Cooking and the Public/Private Split

The public/private split is a feature of most societies, and is particularly marked in those which are hierarchically structured, such as Western societies. The private sphere is associated with the home and women, and so has a low value and status, whilst the public sphere is associated with men and consequently has high value and status. Cooking illustrates in a particularly striking way this unequal and hierarchical relationship between men and women, because the quintessentially female activity of cooking is elevated to a high status when done by men. In order to emphasise this distinction, the development of cooking as a male profession was accompanied by a deliberate distancing by male chefs from everyday domestic cooking. This separation or differentiation became especially marked with the development of French '*haute cuisine*', so that by the nineteenth century, "French professional chefs were at pains to differentiate their work from mere domestic cookery" (Mennell 1985, 201). Although in England there was a less pronounced culinary hierarchy, with women cooks being employed in many well-to-do households, French-trained male chefs were still preferred in upper-class establishments. Typical of the chauvinist attitudes is that of Frenchman, Chatillon-Plessis, who claimed it a 'false economy' to employ a woman cook rather than a male chef because a "woman cook can never give a table the attractive style that a male chef can bring to it (and only a male chef has) the elements of ingenuity which a woman will never know how to carry off...." (201).

Nevertheless, during the second half of the nineteenth century increasing numbers of women were employed as cooks in French middle-class homes, as well as in "smaller and less-celebrated restaurants", a trend that alarmed male chefs who were determined "to emphasis the distance between their own 'grand cuisine' and the ordinary cooking associated with women cooks and housewives" (Mennell 1985, 204). This position was expressed in the policy of the exclusively male union of chefs "not to abstain from giving recipes for simple family dinners, but without descending to what is called *la petite cuisine* or domestic cookery" (204). Sadly, such male chauvinism does not only belong to a bygone age; famous contemporary French chef, Paul Bocuse, despite praising his mother's cooking, nevertheless considered that "like women generally, she lacked the imagination essential to genuine creativity" (Jones 1993, 73). He also infamously contended that women belonged in bed, not in the professional kitchen. 'International superstar' Bocuse, who is credited—and indeed credits himself—with being the creator of *haute cuisine lyonnaise,* in fact trained with la mere Brazier, one of the renowned women cooks and small restaurateurs of Lyons, known as *les meres lyonnaises.* Despite this heritage of women cooks, including his mother, he dedicates his cookbook '*The Cuisine of Paul Bocuse*' (1978) to his father and "in tribute to Alfred Guerot and Ferdinand Point, who inspired *La*

Nouvelle Cuisine Francaise". An example of how famous chefs elaborate on the cooking of their peasant mothers or grandmothers is the case of the famous French dish *crepes suzette*; Charpentier, chef to the royalty, developed a new sauce for the crepes, creating a "grand version of the simple orange sauce made by his peasant foster-mother, who was a splendid cook in spite of her poverty" (Grigson 1983, 269). Another example of a famous male chef's appropriation of women's cooking is that described by another great English food writer, Elizabeth David. She explains how Auguste Escoffier, perhaps the most celebrated of all French chefs, nostalgically turned to the traditional Provencal peasant dishes prepared by his mother in his childhood. Taking one—potatoes and artichoke hearts cooked with olive oil, garlic and thyme—he 'refined' it for the gourmet public of Paris by substituting expensive truffles for the garlic and butter for the olive oil. In a further elaboration, he added a *gigot* of lamb—a dish that for his mother would have been a simple roast on a bed of potatoes prepared for a special occasion or festival—thereby combining and elaborating two traditional Provencal peasant dishes, naming his creation *carre d'agneau Mistral*, in honour of the Provencal poet Frederic Mistral. David wryly remarks that: "He published the recipe in a book: and another of the myriad village dishes of France has entered the repertory of *la cuisine classique*…a poor Provencal family might find their great son's version of a familiar dish lacking in savour" (David 1986, 250), preferring in fact the more robust rustic original to Escoffier's refined version of it.

It was in fact the traditional regional cooking of women, especially that of *les meres lyonnaises,* which was largely responsible for the 'culinary revolution' of the 1970s—that is, the simplifying and lightening up of the rich and over-elaborate '*haute cuisine*' which had been developed by the male chefs of France over several centuries (Kamman 2002). Despite being inspired by the domestic cooking of women, the development of what was known as '*nouvelle cuisine*' was however mediated by male chefs, principally Paul Bocuse and Ferdinand Point, both of whom drew on the cooking of their mothers. Indeed, as Madeleine Kamman says, lamenting the lack of recognition of women's culinary creativity and genius, and the contrasting accolades given to men: "Where are you, my France, where women cooked, where the stars in cooking did not go to men anxious for publicity but to women with worn hands stained by vegetables, peeled and parched by work in house, garden or fields, wrinkled by age and experience" (Kamman 2002, intro.).

The assumption that the cooking of women needs to be 'mediated' by male chefs in order to raise it to the level of art or 'high culture' permeates the thinking of most historians of cooking. Revel, for instance, whilst stating that it is a mistake that, in order to be considered good, cooking has to be complicated and elaborate, nevertheless insists that cooking can only be raised to the level of 'art' when a (presumably male) chef transposes the ingredients "into a new register, where they disappear only to be reborn as a whole that owes its existence to intelligence" (Revel 1982, 13). Revel is very critical of '*haute cuisine*'s tendency to be too elaborate, that such "cooks ruin cuisine: they are the plague of modern gastronomy" (22). He stresses that, in order to avoid the risk of over-elaboration, practitioners of '*haute cuisine*' need to turn at regular intervals to traditional 'popular' (that is, women's domestic) cooking for inspiration. Nevertheless, he warns that "reformers of gastronomy...must know how to react against family cuisine, which clings to its errors as to its qualities" (23). Although conceding that some of the best food he has ever tasted have been amongst the poor peasants of Mexico and Sicily, and that the "more primitive a dish is, the more tasty it is", he considers that "it does not belong to the domain of art, but rather that of ethnology or a mixture of biology and ethnology" (22).

'How Can I Make It My Own?': Male Individualism vs Female Collectivism/Community

One of the distinguishing features of male professional cooking is its individualism. Famous nineteenth-century French chef, Antonin Careme, for instance, considered that his own achievements were the pinnacle of French '*haute cuisine*'. As Revel asserts (uncritically): "Chefs share with philosophers and painters the unquestioned conviction that their time is the only one that is worth anything, and it is summed up, once and for all, in a single man: the one who is speaking" (Revel 1982, 233). One of the distinguishing features of '*haute cuisine*' is its concern, indeed obsession with originality and the 'invention' of new recipes. Whilst acknowledging that the existence of the *haute cuisine* depends on 'popular' cuisine, that is, *haute cuisine*'s debt to the traditions of female domestic cooking, Revel follows the age-old male supremacist dismissal or trivialisation of women's culinary inventiveness and creativity. He adheres to the masculinist tradition of needing to elevate male activity and creativity above that of women, celebrating masculine 'individual originality' whilst deriding the collective tradition of female culinary knowledge: "The gastronome is (not) imprisoned by tradition…With gastronomy cuisine *ceases to be collective*. Everyone prides himself on exercising his original opinion and perhaps contributing his own personal creation" (149, italics added).

The ego-driven, competitive ambition of many male chefs is driven by a desire for fame, a fame which is usually associated with the 'invention' of 'new and original' recipes, but which in fact are often derived from their mother's or other women's regional domestic cooking, as in the case of 'superstar' Paul Bocuse. This individualism contrasts sharply with the collective tradition of female cooks. Speaking of the age-old collective knowledge of women cooks in the south of France, Louise Jones describes how: "Such traditions, such techniques all depend upon a tight network of women, a strong solidarity from place to place and from generation to generation. Asked about her almond tart, Madame Eysette (who cooks in her own small restaurant) would not dream of presenting it as an original creation, but proudly says she got the recipe from Madam Colomb down the road" (Jones 1993, 76). Because women know on some level that they are part of a long female tradition of culinary culture, they don't have the same urge/need to invent a recipe in the way that men do. Therefore, they might do their own version of a traditional recipe, but in keeping with and respecting time-honoured traditions. This is how a culinary culture is developed and sustained. As a result, they don't need to take a radical, revolutionary approach to cooking, such as molecular gastronomy.

Even when male chefs acknowledge the influence of their mother's cooking, "in order to reach their full potential, they have to break away from tradition" (Harris and Giuffre 2015, 60). As Gary Fine points out, male chefs need to emphasise how their cooking is different from the everyday unpaid cooking done by women, with the emphasis on technical skills and 'artistry' being a way of avoiding the threat of feminisation. This "helps to uphold the home-versus-haute dichotomy of professional cooking by casting home cooking as…inferior" (cited in Harris and Giuffre 2015, 51). This denial and avoidance is also a way of shoring up "their sense of precarious masculinity" (41). In their obsession with breaking with/overturning the traditional cooking of their mothers and grandmothers, whilst using it as a reference point, male chefs sometimes go to extraordinary lengths. The phenomenon known as molecular gastronomy is probably the most extreme example of this. Indeed, the desire of male chefs to transpose ingredients into a 'new register' surely reaches its apotheosis in molecular gastronomy.

Molecular Gastronomy as an Expression of the Masculinist Approach to Cooking

Molecular gastronomy was developed during the 1980s by Hungarian physicist Nicholas Kurti, whose hobby was cooking, and French chemist Hervé This, who was "upset by the poor and unscientific way that people cook", and who wanted to scientifically disprove 'old wives tales and tricks' (This 2006) with regard to cooking and food preparation, which in itself is an arrogant dismissal of women's rich historical collective culinary knowledge developed over many generations. Critical of the fact that the basic methods of cooking—such as boiling, frying and baking—had not changed in centuries, Kurti and This were determined to introduce a scientific approach to cooking in which traditional recipes were subjected to scientific scrutiny and experimentation. During the 1990s they developed what is now popularly known as molecular gastronomy in collaboration with chefs, most notably the Catalan chef, Ferran Adrià, the most famous exponent of molecular gastronomy. Accolades and near mythic status are given particularly to male chefs who at some turning point in their career break with traditional cooking, and who, "if talented enough,...end up transforming not only their style of cooking, but also influencing a countless number of chefs who come after them" (Harris and Giuffre 2015, 58). In the case of Adrià, this critical turning point occurred during a visit to French chef, Maximin, who instructed: "It is up to you to create your own cuisine. All the recipes in the (cook) book written by me are exclusively mine. I have created them. It is my cuisine, not that of some other chef" (Andrews 2010, 44). As a result, Adrià was inspired to break with tradition, including *nouvelle cuisine*, and to strike out and create something completely new and his own. The vehicle for this was molecular gastronomy, with one of his early 'creations' being a chicken curry in which the sauce is reconfigured as solid and the chicken as liquid (Harris and Giuffe 2015, 61).

Molecular gastronomy involves the use of an array of processes involving aerosol gases, such as liquefying, freezing and drying, resulting in the creation of emulsions, froths, foams and gels. This is not so much about the preparation of food as nourishment, but about the creation of a 'concept' in which bizarre flavour and unexpected textures and sensations, such as blue cheese and chocolate foam, brown onion ice-cream, Earl Grey tea jelly and liquorice ravioli are designed to surprise the senses. Arguing that "traditional cooking is not a guarantee either for healthy food or for rational preparation of food", Hervé This proposes that molecular gastronomy offers a potential solution to modern food-related problems such as obesity because: "If we are able to use the knowledge gained on food preparation, we might find new ways to make healthy food more attractive, we might persuade more people to cook better food and last but not least, we might convince society to regard eating as a pleasure rather than a necessity" (This 2006) He goes on to claim in this article that he has collected 25,0000 'culinary precisions', a term he prefers to 'recipes', whereby he scientifically tests the reliability of recipes, which he considers are based on 'old wives' tales and tricks'. Surely this represents the epitome of male arrogance, denigration and dismissal of the collective knowledge and wisdom of countless generations of female domestic cooks, whose own experimentation and innovation over the centuries have built up the world's hugely diverse culinary heritage, based on the use of local, seasonal and healthy ingredients. Indeed, in a more recent book, *Note-by-note Cooking: The Future of Food* (2016), Hervé This goes even further, making the extraordinary claim that his new invention will solve world hunger! By "creating dishes with foods that have been deconstructed into basic compounds...(such as) foams and gels, (food) can be transported and rebuilt....bringing nutritious meals to communities around the world, ...(making) cooking more nutritious, energy efficient, and environmentally sustainable" (globalcitizen.org. 2017). This clinical, sterile, disembodied approach to cooking epitomises a masculinist scientific mindset that treats food purely as an assemblage of nutrient elements, divorced from its cultural, geographic and historical context. As a nutritionist, Marion Nestle correctly observes: "Food is not just a collection of chemicals – it has a role in human society that goes way beyond molecules. Eating has social, economic and anthropological roles, and the lack of food is a complicated problem, requiring more than a technical solution" (cited in Ball 2015).

Women's traditional cooking, of which Hervé This is so critical, stands in stark contrast to the clinical and disembodied 'culinary precisions' of molecular gastronomy. In contrast to molecular gastronomy developed in a laboratory, the day-to-day domestic cooking of women, which, as I have elaborated, forms the basis of the world's regional cuisines, has been created and developed from each cuisine's geographic, climatic and historical situation. As such, it carries with it a rich and deeply embodied culinary cultural memory associated with a particular place. A vivid example of this sensory association is described by anthropologist, David Sutton, who during his research into the food culture of the Greek island of Kalymnos was often told "to eat, so that you remember Kalymnos" (Sutton 2014, 8). Cooking does of course have a basis in science—a fact that domestic cooks have probably always intuitively known, such as the physics of temperature and the chemical reactions of different ingredients. Indeed cooking probably formed the original basis of alchemy. However, molecular gastronomy, with its bizarre space-age creations in the form of gels, aerosol foams and emulsions conjured up in clinically sterile laboratories is a quintessentially masculine approach to food and is a far cry from warmth of the domestic kitchen, a place traditionally associated with good nourishing food, warmth and conviviality. It is interesting to note that, when the staff at Ferran Adrià's famed restaurant El Bulli sit down to eat, they tuck into hearty plates of pasta; in other words real food, rather than the artificial, technical creations they devise in their kitchen/laboratory. As, Adrià himself says of the meals the restaurant staff eat: "People are often surprised when we tell them we eat ordinary food. The food we like to eat at El Bulli is the same as what most people like to eat" (Adrià 2011, 7). This highlights the contrast between the nutritional aspects of 'ordinary cooking' and molecular gastronomy, in which the latter is not so much about nourishment, but principally about the need of professional male chefs to stand out as 'different'. This reflects the view expressed by Revel that the fundamental question of gastronomy and *haute cuisine* is "how to eat well without really taking nourishment", and that hunger stands in the way of the true appreciation of food "because at such a time any dish seems delicious" (Revel 1982, 13). Indeed, for male chefs, the nourishment aspect of food is often incidental or subordinate to the display, and even considered to distract from it.

A striking feature of the way in which famous male chefs are spoken about by admirers and the media is the awed and reverential tones. Adrià, for instance is described as a philosopher, a visionary, a genius, a 'god'—even his apparent modesty is a sign of his 'greatness'. One witness to Adrià's approach to food speaks in awed tones of his "always trying something different. I remember one day he opened up a squid with a scalpel and put herbs inside" (Andrews 2010, 161). This does not sound unusually innovative—perhaps it's the use of a scalpel, rather than a knife which is so impressive! The chefs working at El Bulli are "encouraged to play with their food. Everything is fair game, every culture, every ingredient, every conceivable way of dealing with it" (203). Intricate and detailed notes are made of every 'experiment', of which there are about 5,000. This obsessive, verging on fetishist, rarefied treatment of food, which appeals no doubt to well-fed diners looking for new gustatory sensations, is in fact a very masculinist approach to food. Female cooks and chefs, by contrast, are very unlikely to have such an obsessive, single-minded approach to food. Female domestic cooks for a start do not have the time and resources, never mind the inclination, to engage in such obsessive and intricate manipulation of food. Moreover, female chefs do not have the same ego-driven desire to be 'unique' and 'different' which is such a feature of male chef's approach to cooking. Perhaps this is because women know on some instinctive level that they are the inheritors of a centuries and millennia-long female intergenerational culinary heritage that, as I have argued in this chapter, forms the basis of the world's countless regional cuisines. As a result, they do not have the urge or compulsion that male chefs so often have, to break with tradition and create something totally new and unique, that they can claim as 'their own'. Instead female chefs are satisfied with cooking dishes that are faithful to tradition, whilst adding their own particular interpretation, as female cooks have always done. And, frankly, this is the kind of food that most people prefer to eat: a holistic approach to food that is recognisable, delicious and nourishing.

The careers of thirty-three female chefs who have "finally been given equal status with their male (and often highly chauvinistic) competitors" (Pudlowski and Rougemont 2009, 7) is described in '*Great Women Chefs of Europe*'. In contrast to their self-promoting male counterparts, these women "tend to shun the limelight... (and whose) self-effacing modesty and discretion remain their hallmarks" (8). Women also cook to please others rather than to impress. As Deborah Harris and Patti Giuffre found from their research on women chefs: "Unlike men chefs, women chefs place the diner's experience above their own vision or ego", although this in itself is counted against them by the food media as 'proof' that women are less daring and innovative than male chefs (Harris and Giuffre 2015, 58). Another striking difference between male and female chefs is that the former are often concerned with building an empire of restaurants, sometimes on a global scale, whereas women chefs generally are content to focus on doing what they love, cooking to "make people happy….they are not interested in obtaining celebrity status and building empires. They are exactly where they want to be – at the stoves in their kitchens" (78). Women chefs don't crave the media attention that men tend to do, often being uncomfortable with it, preferring to concentrate on their work 'under the radar'. This inclines them to take a more collectivist approach to their cooking, rather than striving to be individual and 'unique', even though this masculine approach is the one that generally gathers the accolades from the food media. The collective nature of women's cooking and, in particular, women's intergenerational food culture is examined in Chapter 5. The competitive and obsessive approach that many male chefs often have is reflected in a high incidence of drug and alcohol abuse, and burnout, as described by Anthony Bourdain in his book '*Kitchen Confidential*' and, even tragically, in suicide, as in the case of the French chef, Bernard Loiseau, who shot himself in 2003 after his restaurant was demoted by Michelin from three stars to two (Chelminski 2005). Indeed, sadly, suicide is not uncommon amongst high-achieving and driven male chefs.

Men Are no Longer Satisfied with 'Simply' Being Chefs

Another difference in women's approach to food and cooking is that it is intricately connected to nourishment. This traditional association between cooking and health is generally disregarded by '*haute cuisine*'. Famous French chef Marie-Antoine Careme (1784–1833) was the first to deliberately break the age-old female tradition of food as having nutritional and medicinal qualities, emphasising "a 'pure gastronomic aesthetic' and the view that food could serve as art for art's sake" (Harris and Giuffre 2015, 22). This approach to food is echoed in Revel's stricture that cuisine should be about "eating well without really taking nourishment" (Revel 1982, 13). Nevertheless, in the constant search for something new and different, some male chefs have turned to a less technical approach to cooking, rediscovering what are essentially female culinary traditions such as the use of wild plants and herbs. Men are no longer satisfied with simply being chefs; they need to be at the vanguard of new food movements, such as 'farm to plate' eating and sustainability. They regard themselves and are regarded by the food media as the 'new philosophers', leading a 'food revolution'. Perhaps the best known exponent of this is the Danish chef, Rene Rendzepi, who is famed for 'discovering' the use of 'weeds' and wild plants in cooking, and whose restaurant Noma has been voted best in the world. There is of course no acknowledgement that the use of wild plants in cooking is something that women have done cross-culturally for millennia, and still do in many less industrialised countries. Male chefs have also 'discovered' seasonal cooking, using locally sourced ingredients. This is celebrated as something revolutionary, although it is what all regional traditional cooking is based upon. And, as Harris and Giuffre point out, this approach to restaurant cooking was first pioneered by Californian chef, Alice Waters in the early 1970s. Despite, or even because of the female culinary legacy that these new male chefs have 'discovered', they have "resisted the linkages of these new cooking styles to femininity. Instead, these forms of cooking have been recast to emphasis masculine traits. Men chefs are featured in food magazines crouching in fields or travelling to exotic locales" (Harris and Giuffre 2015, 40), where they 'cook up a storm', usually outdoors against a dramatic backdrop. Indeed, male chefs commonly visit exotic places precisely to pick up ideas and inspiration from traditional cuisines, after which they return home to re-create in a more 'refined' and 'better' version that they can make their 'own'.

Harris and Giuffre ask why it is that the cooking of female chefs does not receive the same accolades that men are given, with women's cooking being seen as more 'homely' and comforting, more like traditional home cooking as opposed to men's tradition-breaking approach. They seem to suggest that if women were more tradition-breaking they too might receive similar accolades. I would contend, however, that even if they did, they would still not get the same praise as men. The above examples of men 'discovering' traditional female food and culinary skills show that in male supremacist culture men will always have more recognition than women. Harris and Giuffre themselves conclude that numerous studies have found that even when men and women have the same skills, people accord men's skills a higher value than women's (83). This echoes Margaret Mead's perceptive observation, noted in Chapter 2, with regard to traditional women's activities having a higher status when performed by men.

A further example of male chefs claiming to have discovered a 'new' approach to cooking is that of American 'philosopher' chef, Dan Barber. In his book *The Third Plate: Field Notes on the Future of Food*, he describes how, in his search for a sustainable way of eating, he rightly realised that simply replacing industrially farmed meat with free-range organic meat was not sufficient because it still put animal protein at the centre of the plate, with vegetables playing second fiddle. He came to realise that the "right kind of cooking and the right kind of farming are one and the same...and that the very best cuisines – French, Italian, Indian, and Chinese, among others – were built around the idea that the limited offerings of peasant farming meant that grains and vegetables assume centre stage, with a smattering of meat, most often lesser cuts such as neck or shank. Classic dishes emerged – pot-au-feu in French cuisine, polenta in Italian, paella in Spanish – to take advantage ... of what the land could supply" (Barber 2014, 16). Unfortunately, but in keeping with masculinist tradition, Barber fails to mention the countless generations of female domestic cooks who developed this integrated, sustainable way of eating, incorporating what was seasonally and locally available, in the process developing and supporting rich and biologically diverse agricultural systems and regional culinary traditions.

Male chefs are now even promoted, and indeed promote themselves as saviours of humanity, with extreme claims being made that not only are they "the most qualified people on the planet to talk about food, (but) they spur farming best practices to enhance nutritional value, increase efficiency,(and even more ...) *get healthy foods to people in the areas of the world most affected by hunger*" (Global citizen 2017; italics added). The eleven chefs include celebrity names, such as Italian chef, Massimo Bottura, American chefs, Dan Barber and Sam Kass and Anthony Bourdain, French chef, Bruno Loubet, Spanish chef, Joan Roca, the French chemist, Hervé This, and a couple of token women chefs, including Thomasina Miers. Their main method of achieving this is to demonstrate how food waste can be turned into interesting, delicious and healthy meals. No doubt this is a worthy motive, given that in the industrial food system, up to half the food produced is wasted, adding 3.5–4.5% of global greenhouse gas emissions (GRAIN 2016). However, if these chefs, well-meaning though they might be, believe that this approach is going to feed the world, they are both naive and arrogant, and indeed ignorant of the real reasons why over a billion people on the planet are hungry and malnourished. These are to do with many factors, including poverty; the lack of support for small-scale family farming, especially that done by women; land grabbing by multinational corporations; the many negative effects caused by industrial production, such as monocultures, seed patenting and industrial agriculture's total dependence on fossil fuels, resulting in the most serious threat of all—climate change. These are some of the issues I examine in Chapter 7.

Furthermore, once again, these celebrity chefs fail to recognise or acknowledge that the kind of non-wasteful type of cooking that they advocate has been practised by countless generations of frugal and inventive female domestic cooks throughout the world for millennia. Indeed, three-star celebrity chef, Massimo Bottura has published a book *Bread Is Gold* (2017) featuring recipes from other chefs, including those mentioned above, based on left-overs and ingredients such as stale bread. This is of course the very approach to cooking that female domestic cooks have always practised, creating classics out of frugal necessity and developing entire cuisines out of humble ingredients, such as stale bread. However, any acknowledgement of this rich female culinary heritage is, as usual, absent. The real solution to world hunger lies in the recognition and celebration of women's worldwide historical role in the creation and maintenance of nutritious, delicious and sustainable ways of producing and preparing food. It is instructive that during the current COVID-19 pandemic when home cooking has experienced a revival due to restaurants, cafes and pubs being closed, people are turning to the comfort of uncomplicated traditional recipes. No doubt, as men are also cooking more, they are likely to continue with complicated 'chefy' recipes in order to show off their expertise, whilst women in particular are returning to the simple, tasty, comforting and reliable recipes of their mothers and grandmothers. A case in point is that of Australian 1930s cookbook '*Central Cookery Book*' written by Tasmanian woman Alice Irvine, which since the coronavirus, has been the most popular book, according to the owner of one bookshop, with orders for it coming from as far afield as Hawaii (ABC news, 24/5/20). In the following chapter, I discuss and analyse how the different gendered approach to cooking plays out in the context of the domestic kitchen.

Note

1. A modern day example of this 'plundering' of exotic dishes from faraway places is that of chefs, invariably male, who have television shows about them visiting exotic places in search of new ideas, then bringing the recipes back to their restaurant kitchens, where they reinterpret them by elaborating on them to make them 'their own'.

References

Adrià, Ferran. 2011. *The Family Meal: Home Cooking with Ferran Adria*. London and New York: Phaidon.

Alford, Jeffrey, and Naomi Duguid. 2008. *Beyond the Great Wall: Recipes and Travels in the Other China*. New York: Artisan Publishing.

Andrews, Colman. 2010. *Reinventing Food: Ferran Adrià: The Man Who Changed to Way We Eat*. London and New York: Phaidon.

Ball, Aimee Lee. 2015. "Food Matters: Herve This and the Future of Food." *New York Times*, 17 September. https://www.nytimes.com/2015/09/17/t-magazine/Hervé-this-nbn-future-food.html. Downloaded May 26, 2018.

Barber, Dan. 2014. *The Third Plate: Field Notes on the Future of Food*. New York: The Penguin Press.

Bharadwaj, Monisha. 2006. *India's Vegetarian Cooking*. London: Kyle Cathie Limited.

Chelminski, Rudolph. 2005. *The Perfectionist*. London: Penguin Books.

David, Elizabeth. 1986. *An Omelette and a Glass of Wine*. London: Penguin Books.

Ertug, Fusun. 2003. "Gendering the Tradition of Plant Gathering in Central Anatolia (Turkey)." In *Women and Plants: Gender Relations in Biodiversity Management and Conservation*, edited by Patricia L. Howard. 183–196. London and New York: Zed Books.

Field, Carol. 1997. *In Nonna's Kitchen: Recipes and Traditions from Italy's Grandmothers (1997)*. New York: HarperCollins.

Fragner, Bert. 2000. "From the Caucasus to the Roof of the World: A Culinary Adventure". In *A Taste of Thyme: Culinary Cultures of the Middle East*, edited by Sami Zubaida and Richard Tapper, 49–62. London and New York: Taurus Parke Paperbacks.

Frye, Marilyn. 1983. *The Politics of Reality: Essays in Feminist Theory*. Trumansburg, NY: The Crossing Press.

Global Citizen. 2017. "11 Famous Chefs Who Are Trying to Save the World with Food." Accessed July 10, 2019. https://www.globalcitizen.org/en/content/chefs-food-world-hunger-malnutrition-efficiency/.

Goody, Jack. 1982. *Cooking, Class and Cuisine: A Study in Comparative Sociology*. Cambridge: Cambridge University Press.

GRAIN. 2016. *The Great Climate Robbery: How the Food System Drives Climate Change and What We Can Do About It*. New Delhi GRAIN/Daraja Press and Tulika Press; Melbourne: Spinifex Press.

Grigson, Jane. 1983. *The Fruit Book*. Suffolk, UK: Penguin Books.

Harris, Deborah, and Patti Giuffre. 2015. *Taking the Heat: Women Chefs and Gender Inequality in the Professional Kitchen*. New Brunswick, NJ and London: Rutgers University Press.

Hekmat, Forough. 2005. *The Art of Persian Cooking*. New York: Hippocrene Books.

Howard, Patricia L. 2003. *The Major Importance of 'Minor' Resources: Women and Plant Biodiversity*. Gatekeeper Series 112, January. Published by International Institute for Environment and Development.

Hryciuk, Renata E. 2019. *La Alquimista de los Sabores: Gastronomic Heritage, Gender, and the Tourist Imaginary in Mexico*. Institute of Ethnography and Cultural Anthropology, University of Warsaw. https://orcid.org/0000-002-6740-514.

Jones, Louise. 1993. "Women Cooks of Provence." *Convivium: The Journal of Good Eating* 1 (3).

Kamman, Madeleine. 2002. *When French Women Cook: A Gastronomic Memoir*. Berkeley, CA: Ten Speed Press.

Kremezi, Aglia. 2014. *Mediterranean Vegetarian Feasts*. New York: Stewart, Tabori and Chang Inc.

McCann, James. 2010. *Stirring the Pot: A History of African Cuisine*. New York: Hurst and Company.

Mennell, Stephen. 1985. *All Manners of Food: Eating and Taste in England and France from the Middle Ages to the Present*. Oxford: Basil Blackwell.

Mentinis, Mihalis. 2016. *The Psychopolitics of Food: Culinary Rites of Passage in the Neoliberal Age*. London: Routledge.

Mintz, Sydney. 1996. *Tasting Food, Tasting Freedom: Excursions into Eating, Culture and the Past*. Boston: Beacon Press.

Plumwood, Val. 1993. *Feminism and the Mastery of Nature*. London: Routledge.

Pudlowski, Gilles, and Maurice Rougemont. 2009. *Great Women Chefs of Europe*. Paris: Flammanion.

Revel, Jean-Francois. 1982. *Culture and Cuisine: A Journey Through the History of Food*. Translated by Helen R. Lane. New York: Da Capo.

Roden, Claudia. 2014. *The Food of Italy; Region by Region*. London: Square Peg Publishing.

Sutton, David, E. 2014. *Secrets from the Greek Kitchen: Cooking, Skill and Everyday Life on an Aegean Island*. Berkley: University California Press.

Swinbank, Vicki A. 2002. "The Sexual Politics of Cooking: A Feminist Analysis of Culinary Hierarchy in Western Culture." *Journal of Historical Sociology* 15 (4): 464–494.

Symons, Michael. 2004. *A History of Cooks and Cooking*. Chicago: University of Chicago Press.

This, Hervé. 2006. *Food for Tomorrow? How the Scientific Discipline of Molecular Gastronomy Could Change the Way We Eat* 7 (11) (November): 1062–1066. Science and Society, EMBO Reports.

Unsal, Ayfer. 2015. "Cultural Meetings in Istanbul: Ayfer Tuzcu Unsal, or About Mediterranean Cuisine." *YouTube Interview* https://progettomediterranea.com/en/2015/04/18/ayfer-tuzcu-unsal-en/.

Woodward, Sarah. 2002. *The Ottoman Kitchen: Modern Recipes from Turkey, Greece, the Balkans, Lebanon, Syria and Beyond*. New York: Interlink Books.

CHAPTER 4

The Sexual Politics of Domestic Cooking

Cross-culturally women's central role is that of preparing food for their families. Indeed, so pivotal and taken for granted is this role to family life that, as British sociologists Nickie Charles and Marion Kerr put it: "Within the family, woman's place is defined as being in the kitchen, cooking for men and children, feeding them, nourishing them" (Charles and Kerr 1988, 6). In her influential study of this issue, Marjorie DeVault argues that the provision of meals is pivotal to family existence, with a "strong association [existing] between 'mothering' and the preparation of food" (DeVault 1991, 104). Although caring work is not exclusively limited to women, cultural division of labour places far greater responsibility for care on women than it does on men, with perhaps the most important aspect of this caring being in the preparation and provision of meals. Women, unlike men, have been deeply socialised "that care for others is their duty" (119); whereas caring work for men is very much the exception, for women it is obligatory and is considered by both men and women as 'natural'.

From the women she interviewed in her study, DeVault found that as "they grew up, they learned both from their own mothers and from more general ideas of what mothers *should* be, to participate in the social relations that organise the family work of feeding" (96, original italics). She argues that the caring work that women perform in families is central to the construction of gender, and that "feeding work has become one

V. A. Swinbank, *Women's Food Matters*,
https://doi.org/10.1007/978-3-030-70396-7_4

of the primary ways that women 'do' gender...By feeding the family, a woman conducts herself as recognisably womanly" (118). This is a belief that is internalised by women themselves; in her study DeVault found that women considered the providing of appetising and nutritious meals for their families as central to their roles as wives and mothers, even if they felt ambivalent about the work involved. In her study of about thirty married women in South Wales representing a cross-section of socio-economic groups, Anne Murcott (1995) found that all of them "took it for granted that cooking was women's work" (89), whether they enjoyed it or not. All of these women considered cooking to be important for family health and well-being with a strong "equation between proper eating and home cooking" (91), and that it was especially important for the 'man of the house' to come home to a cooked meal.

A central feature of women's daily domestic cooking is the fact that food preferences are overwhelmingly dictated by men, and to a lesser extent by children. In her study of two hundred mothers in the North of England, Nickie Charles found that women's own tastes and preferences were totally subordinated to that of their families, especially to men's, and that part of women's role as carers is the denial of their own needs and preferences, with their "position as 'servers and providers'...involv[ing] a systematic subordination of their own preferences to those of their partners" (Charles 1995, 106). Both Murcott and Charles found that men's preferences invariably took precedence, and that "it remains, however irksome, up to the woman to prepare what a man wants" (Murcott 1995, 97). Even when women expressed enjoyment and a sense of autonomy in their cooking, this was still subject to men's tastes and demands. DeVault argues that women's deference to men's tastes and preferences is a result of the socialisation of women to accommodate and subordinate their own needs and preferences to others, especially to those of their husbands. Resistance to this role is considered selfish and the antithesis of women's caring role which is so central to the ideology of the family. As DeVault puts it, "[S]ince feeding work is associated so strongly with women's love and caring for their families"...if they resist, they..."risk the charge – not only from others, but in their own minds as well – that they do not care for the family" (DeVault 1991, 154).

Whilst women often insist that such accommodation to others is a result of choice on their part rather than deference, such 'choices' are often made in order to avoid conflict. Indeed, DeVault found from her survey of thirty households that there was remarkably little conflict over housework, but, rather than this being an expression of equity, the relative lack of conflict was due to women's accommodation to others, this being "preferable to sustained conflict" (163). Whereas women often express gratitude if their husband is 'easy to please', Charles points out that the refusal by some men to eat the food that is served to them is very upsetting for women, and reinforces their catering to men's preferences in order to try and avoid this.

Food and Male Control

Women's accommodation to men's preferences as an attempt to avoid unpleasantness and conflict is not always successful. Issues around food and meals are frequently a feature of marital or domestic violence. Rhian Ellis reports that nearly all research on domestic and marital violence "has data on battered women who say their husbands attacked them for failure to have a hot meal ready" (Ellis 1983, 166). The issue of food and meals is prominent in violent relationships because the preparation of meals is central to women's servicing role towards their male partners, and abusive men punish women who are perceived not to be adequately fulfilling that role. Ellis describes, for instance, situations where the woman is beaten for not being up at 2 or 3 am when her husband returns home drunk demanding a hot meal. Such men also maintain tight control of finances, with women not being given sufficient money for food and then being beaten for not being able to buy the steak he demands; of women and children going without so he can have his nightly meal of steak and chips; of women scarcely being allowed out of the house to do the shopping, of having to account for every minute and every penny. Citing a study on domestic violence by Dobash and Dobash, Ellis states that from a young age "the girl is taught to combine deference to the authority of those males to be served", and that such service is "often and centrally concerned with food" (165). Another way in which some men exercise control is by being critical of their spouse's or partner's cooking and refusing to eat it. One shocking example of this kind of abuse was the refusal by Charles Saatchi, the then husband of famous television cook and cookbook writer, Nigella Lawson to eat her food. He preferred Weetbix and processed food 'out of boxes' to her own famously delicious meals, telling her that "there's nothing you can cook me that compares with Weetbix". In a BBC interview in 2011 Lawson recounts that on one occasion, after she had cooked a delicious prawn *dhansak*, he told her that it was “the most disgusting thing I have ever eaten”. Such behaviour would suggest that he resented her huge success and popularity and that his intention was to undermine her.

The family meal is an important means of socialising children, teaching them about social skills such as sharing, communication and table manners. The sharing of meals is generally considered by anthropologists and sociologists to be central to the socialisation of human beings, and indeed, there is much angst about modern eating patterns—‘grazing’, snacking, not sitting down at the family table for meals—and the impact of this on both nutrition and family cohesion. Shared meals—which for most people means family meals—are also considered important from the point of view of nutrition, with research cited by British food writer Joanna Blythman indicating that “families that eat together tend to eat better food ...[and that] regular family meals are related to healthier dietary patterns and higher intakes of fruit and vegetables in older children” (Blythman 2006, 88). Blythman reports that a 2004 survey of British meal habits found that one in five families ate together only once a week or even less, with 75% of families watching television as they ate (compared to 75% of French people who eat regularly at the family table). As a result of this decline, the stock-cube company Oxo dropped its long-standing advertisement featuring a family sitting down to a traditional Sunday roast, as this image no longer conformed to the reality of most people’s lives (85–86).

The family meal is often viewed as an ideal in which communal eating and conversation take place in a harmonious atmosphere. However, the reality is often very different, with the family meal being in fact an arena of conflict, even of potential violence where men assert their control and authority, especially over children, with women striving to keep the peace by avoiding difficult topics of conversation, whilst seeking to keep the children from upsetting their father. Mealtimes are often a forum for patriarchal social control, for men to exert their dominance and authority, especially over children. Nickie Charles points out that men tend to be much more authoritarian than women in their approach to this, often “exert[ing] their authority through the children’s enforced consumption of the proper family meal...with their views prevail[ing]..., even when women may not agree with their insistence that children eat what they are given” (Charles 1995, 112).. She states that often young children are resistant to conforming to strictures of family meals, with women tending to be more patient and tolerant of this than men.

Perhaps most significantly, meals socialise children into the family ideology of gendered power relationships based on male power and authority and female service and subordination:

> Ensuring that children eat proper meals is clearly not only about ensuring that they are well-nourished. It is also part of the process of socialisation through which they learn not only behaviour that is socially acceptable but also about the social relations which constitute families. These relations are unequal in terms of power and the allocation of tasks, both power and tasks being differentiated along lines of gender and age. In consuming meals children are acquiescing to the social relations within which they are consumed, and in teaching children to eat this way women and men are teaching them about the social order. (Nickie Charles 1995, 113)

This perhaps is one reason why there is such concern about the so-called decline of the family meal as it is seen as reducing the opportunity for asserting male authority and instilling into children hetero-patriarchal power relations.

Ruth Shwartz Cowan contends that a serious widely held misconception—including by sociologists and also across the political spectrum from conservatives to Marxists—is "that twentieth century technology has radically transformed the American household….[from] a unit of production to a unit of consumption" (Cowan 1983, 71), the thinking being that whereas once people produced much of their own foods and clothing, they now purchase it from the marketplace. She argues that modern household technology, despite popular thinking that it has decreased housework, has in fact increased women's domestic workload, and has also actually increased the productivity of the average housewife. Whilst it is true that goods are no longer produced in the household for sale in the marketplace as they were prior to industrialisation, the household "continue[s] to produce goods and services intended for use in the home…[and] the household continues to be the locale in which meals, clean laundry, healthy children, and well-fed adults are 'produced'" (Cowan 1983, 100). Speaking of the work involved in 'feeding the family'—work which entails not only the actual physical preparation and cooking of meals, but also all the invisible emotional and thought-work, the planning, organising and catering to different needs—DeVault insists that such work produces family and household life. This kind of work is essential to "the work of feeding a family", but the taken for grantedness and the "invisibility of this thought work, along with the way it combines with physical tasks, can hide this part of housework even from those who do it" (DeVault 1991, 55).

Men's Involvement in Domestic Cooking

Over the last two or three decades, much feminist research and analysis has been done on the domestic division of labour. Study after study has found that men's involvement in domestic work has scarcely increased during this time, despite the increased participation of women in paid work outside the home (e.g. Van Berkel and De Graaf, Kemmer, Besen-Cassino). However, whilst men's increased participation in domestic labour has been negligible, the two areas in which some men have become more involved are shopping and cooking—significantly, the two areas of household work that women themselves have generally found the most enjoyable, or at least the least unpleasant. In her analysis of American cooking and eating habits since the 1950s, Laura Shapiro observes that even though women "still prepare most of the family meals, cooking has become the first domestic chore to float free of sex roles" (Shapiro 2005, 253). Men's domestic cooking, however, rarely involves doing the routine day-to-day family cooking; this is still overwhelmingly done by women, with men preferring to spend leisure time cooking complicated meals for a special occasion. The daily responsibility of preparing family meals is not optional for women, whereas it is for most men—that is, with the rare and notable exception, men cook only when they choose to. This was confirmed by interviews with men and women conducted by Frances Short who found that it was "only [the] men who could see cooking purely as a hobby" (Short 2006).

Men's selectiveness with regard to housework is confirmed by a number of studies on the subject. One such study carried out during the 1990s which looked at the domestic division of labour in over one thousand households "representing a cross-section of the adult Dutch population", and which sought "to show the relevance of discriminating between different sorts of housework", confirmed that when men do housework they make "at best, selective inroads,…[so] that change is limited to the more enjoyable work" (Van Berkel and Graaf 1999, 786). In their study, Michel Van Berkel and Nan Dirk De Graaf found that housework was generally disliked by both sexes, with cleaning, especially of toilets and bathrooms, and bed-making especially disliked, and, therefore, hardly ever done by men. Their survey found that women did 79–83% of the cleaning, and that when domestic help was employed, it was usually for the least enjoyable work, namely cleaning. Their study also found that, whilst wives also did the bulk (71%) of the shopping and cooking, this was the area in which there was the greatest equality, with about 20% of couples sharing cooking and shopping evenly, supporting their view that the "participation of husbands is… not only generally low, [but] is also *clearly more orientated towards relatively pleasant tasks…*" (Van Berkel and Graaf 1999, 798; italics added).

A particularly interesting aspect of Van Berkel and De Graaf's study is the effect of education on the selective division of household labour. They found that when husbands are more highly educated than their wives, the latter do all the housework. When the woman's education level is higher than her partner's, the man's contribution to housework is markedly less than when both are highly educated: "it is noticeable that highly educated wives whose husbands have a low level of education tend to do eighty-four per cent of the cleaning" (Van Berkel and Graaf 1999, 802), compared to 70% amongst couples where both are highly educated. This resistance to household work by men who are less educated than their wives is no doubt due to their resentment at taking on the low-status domestic role associated with women, and in order not to feel 'emasculated', insist on the traditional servicing roles of their wives. This supports the view of DeVault that "housework not only produces goods and services but also reproduces gender relations" (787).

A significant finding of Van Berkel and De Graaf's study was that in more highly educated couples, especially those espousing egalitarianism, the tendency of men to select the pleasanter tasks, such as cooking, was particularly marked. In her study of twenty-two newly married or co-habiting childless couples in Scotland, in which both partners were in full-time study or employment, Debbie Kemmer found that almost half "did not conform to the 'traditional' division of labour in relation to food preparation" (Kemmer 1999, 573). She observed that "for the majority of couples, most of whom had similarly onerous paid work, food preparation was the responsibility of the woman" in addition to most of the other domestic work. However, in a 'very sizeable minority' of the couples she surveyed, the cooking was more or less equally shared, whilst in two, the evening meal was mostly prepared by the men who "both loved to cook and saw it as a hobby rather than a chore" (575). In both of these cases, the women did not enjoy cooking and were extremely appreciative of their partner's involvement in cooking to the extent that they felt apologetic about their lack of cooking. They felt a need to compensate for this by doing "most or nearly all the other housework, including in one case, clearing up after her husband had cooked, and maintaining a clean kitchen for him to work in" (575). This gratitude on the part of women when their male partner cooks, and their need to compensate by doing the rest of the housework, stands in stark contrast to studies that have found that "[W]hen the bulk of the cooking was done by women, so, in most cases, was *the majority of the rest of the housework*" (574; italics added). Kemmer's study supports Van Berkel and De Graaf's findings that in couples that consider themselves egalitarian the men tend to be more involved in cooking, although not in other household tasks. In Kemmer's study, one of the men who did most of the cooking because he "took great pleasure in expanding his culinary horizons", remarked that he considered that he was breaking with the 'sexist role' of the man expecting to have dinner ready, such as had been the case with his parents. Nevertheless, despite this claim to be rejecting sexist stereotyping:

> These couples have certainly not reversed the 'traditional' domestic roles: the men cook *because they enjoy it* and the women perceive themselves as very privileged, and *feel a need to compensate by a greater contribution to the general domestic labour.* (Kemmer 1999, 575, italics added)

It might be assumed that men's interest in cooking is a relatively recent phenomenon, a by-product of the women's liberation movement and a desire for more 'egalitarian' heterosexual relationships. Whilst it is true that this has been a motivating force in some relationships (as illustrated by the studies referred to above), what stands out in many of these 'egalitarian' relationships is the extent to which the domestic division of labour is skewed in men's favour, with men tending to 'cherry-pick' the more rewarding and creative aspects of housework, especially cooking. A study of dual income heterosexual American couples, confirmed that men choose cooking over other housework, such as cleaning, because cooking "doesn't threaten their gender identity in the same way (as cleaning)" (Besen-Cassino and Cassino 2014, 46). Women still overwhelmingly carry the burden of the far less enjoyable jobs such as cleaning, tidying up, washing, unless they can afford to pay someone else—often a poor immigrant woman—to carry out this work. Men often justify their supposed 'inability' to clean as somehow innate to their sex, implying that women are by nature more concerned with cleanliness, more fussy and have a better eye for detail. This essentialist view does not, however, extend to cooking, unless perhaps it is to do with the minutiae and organising of day-to-day cooking. Indeed, when it comes to cooking, essentialism also holds sway in that, despite cooking being considered central to women's caring and nurturing role in the family, there is a long tradition of men asserting that they are by nature superior cooks to women.

Cookbooks for Men

In her study of the influence of cookbooks in the United States during the twentieth century, Jessamyn Neuhaus (2003) found that articles and books on cooking during the 1920s and 1930s emphasised the importance for women of learning to cook well if they were to be successful in finding a husband, and, even more importantly, keeping him and having a family. One typical book of the period written for young married working women described cooking to please one's husband as 'home insurance': "We don't guarantee that reading this book will get you a husband in six weeks, but we do claim that it will help you to keep the one you already have" (Neuhaus 2003, 71). Whilst the vast majority of cookbooks were written for women, especially white middle-class women who could no longer rely on servants and other domestic help as they had previously, and who needed to "assume responsibility for the family's daily meals and…learn to enjoy it" (72), there was also a surprising number of cookbooks written during this period specifically for men.

A striking feature of most of these books, mostly written by men for men (although some were also written by women), was their insistence on what they considered men's innate superiority in the kitchen. Neuhaus points out that cookbooks written for men featured very different kinds of foods than those in women's cookbooks, with an emphasis on 'manly meals' such as steak and hearty stews and strongly flavoured spicy dishes such as chilli in the former, and 'dainty', 'feminine', bland food in the latter. These books also invariably stressed that, whereas women needed to be taught the potential for cooking to be a creative activity, men instinctively knew this, and unlike women, had the natural ability and imagination to raise it to an art form. According to one writer, male cooks "prove the old adage that a man can do anything better than a woman can…they cook rings around their wives" (cited in Neuhaus

2003, 91). Moreover, whereas the assumption of women's cookbooks was that women did the routine day-to-day cooking, men's cookbooks were concerned with cooking as a hobby, a leisure pursuit. In one book cited by Neuhaus, titled '*Men, Meet the Kitchen*', one male author likened men's cooking to a leisure activity such as sport. Like many others writing on this topic, this author was scathing about women's cooking abilities, asserting that men are innately and instinctively better cooks, not even in need of any training; according to him, whereas woman slavishly and unimaginatively followed recipes to the letter, men had the natural flair and creative imagination to create masterpieces with the minimum of guidance. In contrast: "A woman will take a recipe, follow it faithfully and religiously to the dot, and set out a dish which, albeit appetising, is exactly the same dish a thousand other women are setting out at the moment" (cited in Neuhaus 2003, 86).

Men, on the other hand, are much more adventurous, not being "content to let it stop at that. What woman would have thought of adding a cupful of chilli to that old standard salmon soufflé recipe ..., as I did, and thereby achieve a most delicious and novel dish?" (38)—a bizarre combination which sounds less than appealing, and suggests that the vanity of this author lacked any sense of discernment. Neuhaus describes how it was important for men to distinguish their 'manly' cooking from the female domestic sphere: "They had to police diligently the boundaries between kitchen as women's space and kitchen as male hobby space" (94). This distinction of cooking as an occasional leisure activity for men in contrast to women's everyday cooking helped to define and reinforce "the white middle-class woman's daily, domestic role in the twentieth century" (Neuhaus 2003, 76). Laura Shapiro (2005) describes how, especially in the post-Second World War years with the increase in processed, packaged foods aimed at women, the stereotyped distinction between women's and men's cooking became particularly marked with pre-packaged cooking being associated with women's day-to-day cooking, and elaborate 'gourmet' meals prepared from scratch being associated with men (Shapiro 2005, 174). This distinction between men's and women's cooking was further reinforced during the post-war years when there was great pressure on women to return to the domestic sphere. As Neuhaus explains, during the 1950s, as in the 1920s and 1930s, cookbooks and articles by men for men "consistently emphasised that men naturally possessed the ability to cook more creatively and with better results than women" (Neuhaus 2003, 202). In fact, "the 1950's saw an increase in such cookbook rhetoric. Even more regularly than in the pre-war period, cookbooks for and by men insisted not only that men made inspired and excellent cooks, but that women prepared dull and tasteless meals" (204). A rare male writer might suggest that being responsible for the unavoidable day-to-day cooking made it difficult for women to be always inspired in the kitchen and if men do "approach cooking in an adventuresome spirit, [it is] simply because they are not obliged to cook day in day out. It is to them first a game, then a pastime, and finally a hobby " (206).

Sherrie Innes, in her study of cookbooks for men by men during a similar period, also found that these books emphasised men's supposed innate culinary ability. Writer after writer expressed themselves in similar terms:

> When man really sets himself to cooking…he can outdo a woman any day…Generally speaking, men are more original and individual cooks than women. They have a knack for putting unexpected upper-cuts in ordinary dishes and making them really knockout fare. (cited in Innes 2001, 32)

Innes describes what she calls the 'male cooking mystique' that was promoted by and reflected in cookbooks of this period. This consisted of 'masculine' foods considered suitable for men, such as steaks and outdoor barbequing; the insistence that their tastes were more sophisticated and therefore superior to women's and so should always dominate; their claim to an innate feel for cooking which raised it from the mundane to an artistic expression, whilst simultaneously disparaging women's abilities. She contends that this mystique arose from men's need to distance themselves from the 'feminising' influence of domestic cooking, and to trivialise women's domestic work whilst emphasising the importance of men's activities in the public domain. Whilst this is undoubtedly so, something more is also going on here—namely, that in insisting that women are naturally inferior cooks to men, in denying women the imagination and creativity that are involved in cooking, they are denying not just the culinary talents and knowledge of individual women but, even more significantly, women's collective culinary culture. In short, it is an attack on women's culture itself, given that it is women who have collectively developed over many generations the countless cuisines of the world, as I elaborated in Chapter 2. This is in fact an example of male supremacist erasure of women's contribution to human culture whereby women's achievements are ignored and denied, and with the history of human achievement being defined, as feminist philosopher Marilyn Frye puts it, "as a history of the acts and organisations of men", and in which "[T]he phallocratic scheme does not admit women as authors of perception, as seers" (Frye 1983, 163, 165). It would appear that men cannot allow that something as important and as vital to human culture as culinary knowledge and expertise is fundamentally a female domain. As I discussed in Chapter 3, the masculinist need to denigrate women's skills and creativity is also expressed in the creation of a culinary hierarchy in which male chefs' cooking is elevated in status vis-à-vis women's domestic cooking. This denigration of women's cooking is vividly illustrated by the following misogynist view expressed by a male writer during the 1920s, presumably due to his hostility to the suffragette movement:

> As women were born without any real taste for food ...they must learn, now that they are legally declared to be the equals of man, to discover why he has always been superior in that art which for centuries they have unrighteously proclaimed as entirely their own. (cited in Innes 2001, 30)

The fact that (some) men recognise the true importance of domestic cooking, even whilst simultaneously disparaging it, helps explain the fact that this is the one area of domestic life that they choose to involve themselves in. As Mary Drake McFeely (2000) puts it, even though domestic cooking is still seen as primarily women's work, the fact that "more and more men [are] voluntarily step[ping] up to the stove...might be seen by women as a clue to an often unrecognised value of a female art" (3).

Whilst most of the writers who contributed to the myth of the 'natural' superiority of men as cooks were themselves male, there were also some women cookery writers, who, no doubt feeling compelled to flatter men's culinary egos, also ascribed to this mythology. However, there were others who had little patience for this. Both Innes and Neuhaus point out that there were women writers who objected strongly to the male presumption of superiority in the kitchen and the invasion of their kitchens by male 'hobby cooks'. According to one female writer: "In all probability there is not one man in ten thousand, despite his eloquence to the contrary, who can go into a kitchen and produce a dinner for two that has balance and flavor and attractiveness" (cited in Innes 2001, 32). In a 1949 magazine article, another woman referred to male culinary presumptions, as well as their lack of acknowledgement of all the invisible planning work involved in preparing a meal:

> You've all met him. One-dish Harry,... who claims that men are better cooks than women, basing his argument on his solitary talent (of a particular salad or dessert)...The fact that most women prepare entire meals not only on grand occasions, but night after night, he considers not at all...The average male cook, faced with the triple threat of marketing, preparing, and serving an entire meal, would make for the nearest hamburger heaven on the double-quick. (cited in Neuhaus 2003, 210)

Other female writers, in articles with titles such as "Husband, Stay Away from My (Kitchen) Door" and "Whose Kitchen is it Anyway?", complained about the mess men create when they make their 'masterpieces', using every pot and pan in the kitchen and then expecting their wives to clean up after them. Writing about her husband's "self-satisfied salad-making technique", one woman describes how he used every spice and condiment in the kitchen for the salad dressing and all her towels to dry each leaf, leaving behind "a mountainous pile of tangled vegetable matter on the kitchen drainboard". Despite the fact that she has prepared the main course, vegetables and dessert, he "is convinced that he has made dinner" (cited in Neuhaus 2003, 211).

The Rise of the 'Manly' Cook

Neuhaus argues that the need to distinguish men's tastes and appetites from women's was heightened during the 1950s, due to anxieties about a perceived undermining of especially white middle-class men's masculinity as a result of suburban living. During this period when men were increasingly doing sedentary white-collar jobs that didn't require the physical stamina of outdoor work, "[M]anhood in the era of suburban life seemed in need of redefinition" (Neuhaus 2003, 216). These concerns about the 'feminising' influence of suburban life on men led to increased anxiety about the so-called 'homosexual menace' of the 1950s McCarthy era; as a result of such anxieties, the rhetoric of the previous couple of decades which sought to distinguish men's appetites and cooking abilities from women's intensified during the 1950s (Neuhaus 2003). Male anxieties during the 1950s and 1960s about both the 'feminised' urban domestic sphere and middle-class women's increasing involvement in paid work outside the home—what one misogynist observer referred to as "the 'womanisation' of masculine public spaces"—prompted a response to "reclaim the indoors for men" (Hollows 2002, 144). In her study of the food pages that appeared in 'Playboy' magazine during the 1950s and early 1960s, Joanne Hollows identified an attempt to "not only masculinise cooking and food but also to show how it is integral to a masculinised and urban domestic culture which seeks to challenge a feminised and suburban domestic ideal" (Hollows 2002, 143). According to Hollows, 'Playboy' magazine presented an 'alternative masculinity' to that of the suburban male breadwinner, a stereotype which it considered represented both conformity and femininity. The suburban male breadwinner role was replaced by "an alternative middle-class masculinity based on a distinctively urban 'good life' of pleasurable consumption" (144). 'Playboy' sought to appropriate the desirable aspects of domesticity, such as comfort and pleasure, whilst repudiating any associations with femininity. Central to this construction of the urbane middle-class single male was the bachelor apartment in which "the 'manization' of the kitchen – symbol of feminine domesticity – [was] a reflection of the playboy's 'urbane personality'" (145). This was expressed by the owner of 'Playboy' magazine, Hugh Heffner: "In an increasingly womanised society, the domain of the chef – male since time immemorial – has been invaded en masse by chintz-aproned housewives. But today's well-rounded urban male, safe in his bachelor bastion, with The Playboy Gourmet in hand, can still wield the spatula and twirl the swizzle stick with style and dash" (Hugh Heffner, cited in Hollows 2002, 145).

The repudiation of a 'feminised Other' was central to the construction of the playboy lifestyle. The playboy construction of cooking as a masculine activity, in opposition to a feminine one, involved features that were associated with masculinity, including "making very explicit connections between food, cooking and sex" in which cooking is used as a means of seduction and the interest in certain foods purported to have aphrodisiac qualities. Another feature of the 'masculine' cooking promoted by 'Playboy' was an emphasis on extravagant and conspicuous consumption, in contrast to women's cookbooks that stressed the need to carefully budget and economise. As Hollows puts it, "the ability to consume is central to the construction of Playboy's bachelor…the ability to spend on self, rather than the family, is…integral to the playboy lifestyle" (Hollows 2002, 149). Playboy cooking was intended to be an elaborate performance designed to impress others. The playboy cook pursues cooking as an occasional leisure activity, using expensive and extravagant ingredients, and above all, cooks to please himself, and to impress others. This stands in stark contrast to women's domestic cooking and the caring work this involves which is done for others on a daily basis, deferring to the needs and tastes of others and operating under the constraints of cost and time.

The Rise of the Television (Male) Celebrity Chef and the Creation of the Masculine Domestic Cook

Although there have long been male cooking 'experts' on television, it was really only during the 1990s that television screens were deluged with a wave of, usually male, celebrity chefs, epitomised by Jamie Oliver. As Joanna Blythman (2006) points out, until then the most popular and successful television cooking shows in Britain had been those presented by Delia Smith whose "encouraging, straightforward, business-like and…(un)show-offy" manner appealed to "a strongly female following who found her recipes realistic and do-able" (7–8). However, by the late 1990s cooking for cooking's sake was being displaced by the concept of food and cooking as a statement of fashion and lifestyle. As has already been discussed above, this approach to cooking "is more commonly associated with men who tend to treat cooking as a hobby and turn their meals into 'special events'" (Ashley et al. 2004, 183). This approach to cooking as essentially a leisure and lifestyle statement with its promotion of high-profile celebrity, usually male chefs was about food as entertainment, in which cooking was secondary to the TV chef's persona and lifestyle, and as such was intended "to appeal to people with little or no interest in food, people who quite possibly had no intention to cook" (Blythman 2006, 9). In contrast, Delia Smith's educational approach was now considered old-fashioned, rather dull, and "out of touch because of its patronage of the unglamorous, primarily female realm of home cooking" (8). The criticism of Delia Smith's approach was also fuelled by the hostility of certain male celebrity chefs who no doubt resented Smith's massive success as a television cook and her enormous wealth resulting from this. Gary Rhodes, for instance, stated that "being shown how to boil water was insulting to (people's) intelligence", and Antony Worrall Thompson dubbed her "the 'Volvo of cooking' because he considered that she was reliable but dull" (Blythman 2006, 10).

As for Delia herself, she retired in 2003, expressing dismay at the 'infotainment' direction that television cooking shows had taken. This shift in television cooking programmes from practical instruction in cooking skills and know-how, to food as a 'spectator sport', and what Blythman calls 'added-value' cookery shows in which something else—"sex, travel, eccentricity, adventure, farce, incessant swearing – anything"—was needed to sell the food and cooking has resulted in:

> [t]he balance of culinary power in Britain ...[swinging] away from the domestic zone, where its keepers were women who passed on their accumulated knowledge with their egos held well in check, to the male zone, where its new luminaries are a bunch of flashy performance artists...[t]he success of [whom] is not measured by whether the woman along the road successfully cooked one of their dishes, but by their ratings and book sales. (Blythman 2006, 11)

Hollows contends that the lifestyle being promoted by 'Playboy' during the 1950s and 1960s was about the production of both a "masculinized cooking practice" as well as "the production of cooking practices as 'aestheticized leisure' for the new middle classes and their consumer lifestyle" (Hollows 2002, 151). In another article in which she examines the 'domestic masculinity' of British television chef, Jamie Oliver, Hollows argues that the presentation of cooking for leisure as part of a new middle-class lifestyle is not gender-neutral, but in fact "reaffirms masculine domestic cookery as a leisure and leisured activity" because it "presume[s] a position of distance from domestic labour that is more readily available to men than to women" (231). She contends that Oliver's approach to cooking as a 'fun' leisure and lifestyle activity epitomises the "construction of the masculine domestic cook...[which] disavows the extent to which cooking is a form of labour" (229). Hollows further explains:

> Jamie's culinary masculinity, in which domestic cooking is experienced as a form of creative leisure, is also a product of a distance from domestic obligation and labour (and the accompanying experience of time poverty and the need for constant temporal management) associated with women's position in the sexual division of labour. [While] Jamie's dispositions may be open to specific women in specific social and economic conditions,...his position in the sexual division of labour is associated with masculinity. (Hollows 2003, 243)

Hollows points out that, unlike most British male television culinary personalities, Oliver rejects the label of 'chef' by deliberately distancing himself from '*haute cuisine*' and what he refers to as 'chefy' cooking. Oliver, probably more than any other male TV chef, appropriates the culinary female domestic sphere. Nevertheless, in order to avoid the stigma of femininity associated with this sphere, Oliver needs to establish that domestic cooking doesn't threaten his masculinity. He does this by drawing on images of the post-modern 'new man' in order "to construct domestic cooking as recognisably 'manly'" (Hollows 2003, 232), reassuring other men that cooking is a 'cool' masculine activity. Despite pursuing an activity associated with femininity, his cultivation of 'ladishness' enables him to "transgress 'everyday' feminine domestic cookery [and] serves to distinguish him from the 'home-cooking' he appears to embrace; he occupies the domestic but is not contained or defined by it" (242). Indeed, in order to emphasise that his cooking is distinct from that of the female domestic cook, and that his cooking exemplifies 'manliness', Oliver refers, for instance, to having a night out drinking with the boys, followed by eating an "extra-large lamb-kebab", and how a baker's advice to treat bread dough like a woman, "improved (his) bread-making as well as (his) sex life!" (cited in Johnston et al. 2014, 17).

Another way in which Oliver's 'domestic' cooking avoids the stigma of femininity is the way in which, even though he occupies traditional female space, namely the domestic kitchen, he is not confined to it or trapped by it: "Jamie's domestic masculinity is based on an ability to enjoy the feminine pleasures of the domestic sphere as a realm of leisure because it is predicated on the ability to escape" (Hollows 2003, 242). Oliver, in fact, exemplifies 'new age' men's tendency to 'cherry-pick' the most enjoyable and creative aspects of the domestic sphere. In his performance of domestic cooking interspersed with his shopping forays on his Vespa scooter to markets and speciality food shops, he is occupying and enjoying the best of both the private and the public worlds. In his ability to do only those domestic tasks which he finds enjoyable and creative (whilst in addition making a great deal of money from it, plus receiving enormous ego-boosting accolades), Oliver is appropriating and capitalising on a female domestic culinary tradition, whilst simultaneously perpetuating a male tradition of trivialising and negating female domestic cooking in order to inflate and legitimise his own culinary activities. This is because:

> Without a sense of distinction from the domestic cook, professional chefs not only lose their claims to legitimacy and expertise in culinary matters, but are also threatened by the feminine associations of domestic cookery which, for most women, is also ultimately, work. (Ashley et al. 2004, 185)

Jamie Oliver appropriates female domestic culinary tradition without acknowledging it. In order to prove that he is better at domestic cooking than women, he "simultaneously embraces the simple pleasures of domestic cooking, while refusing their 'simplicity' by doing them (in his words) 'really well' and making it 'special',... [thereby] demonstrating that he can both occupy the realm of everyday feminine domestic cookery while rising above it" (Hollows 2003, 242). Indeed, like his numerous predecessors referred to earlier in this chapter, he goes to considerable lengths to assert that his domestic cooking is superior to that of women, making fun of the supposed culinary inadequacy of both his wife and mother-in-law. In a couple of YouTube videos (2012, 2010) Oliver describes his wife Jool's cooking (who he refers to as 'the missus') as 'hideous', and to be avoided if possible. This attitude is part of a long misogynist tradition that belittles women's cooking and recalls a similar one expressed more than half a century earlier, where one man described his recipe for strawberry pie as "not the kind Mother used to make, but the kind Mother used to try to make and wonder why she couldn't" (Neuhaus 2003, 204). Oliver's scarcely concealed resentment at female prowess in the kitchen is evident in his cooking series *Jamie's Italy*, in which he travels around Italy 'showing off' his culinary skills to the locals; however, his cooking is often found wanting because the culinary benchmark for most Italians is their mother's or grandmother's cooking. His barely concealed hostility to this and his underlying sexism is revealed when he has the arrogance to set himself up against a group of village women in a pasta making competition, and when he loses, condescendingly calls one middle-aged woman 'grandma', much to her obvious annoyance.

Television celebrity chefs, such as Jamie Oliver and Yotam Ottolenghi follow the age-old male tradition of seeking out the domestic cooking of women—in Oliver's case that of Italy, and in Ottolenghi's, the Mediterranean—as a source of inspiration and culinary know-how, after which they reinterpret and elaborate on it to make their 'own'. In his cookbook, *Jamie Cooks Italy*, featuring recipes he and his team spent many months collecting from the *nonne*, the grandmothers of Italy, Oliver concedes that they hold the key to the best of Italian regional cuisine, as well as having a wealth of traditional knowledge with regard to sustainability, such as avoiding food waste. However, despite his obvious respect for their great culinary knowledge and skills, he still refers condescendingly to these women, many of them in the eighties and even nineties, as 'these lovely ladies' and 'these girls'. Although the book (and television series on which it is based) pays tribute to the culinary legacy of these women, the book is nevertheless, in true masculinist tradition, dedicated to his two male food mentors and heroes, collaborator Gennaro Contaldo and the late Antonio Carluccio (who was a rare exception, in his genuine respect for the female domestic cooks of his home country). Indeed, the book features a number of photos celebrating male bonding, including a suntanned sporty-looking Jamie on a motor-bike with the much older Contaldo riding pillion, against a picturesque Italian background; in doing so, Oliver confirms his masculine 'laddish' identity so pivotal to his popular image. Whilst Oliver is in awe of these *nonne* with their vast collective reservoir of food knowledge, he clearly also sees it as an opportunity to revive his failing restaurant empire by incorporating some of these *nonne's* recipes. No doubt in their innocence and generosity these old women would be flattered to have their recipes featured in such a way, but the fact remains that this is yet one more example of male appropriation of women's culinary heritage for personal gain.

Another example of Oliver's unacknowledged appropriation of women's ideas and creativity is the case of his well-known (but ultimately unsuccessful) campaign to introduce healthy school meals. This campaign, with which Oliver is so associated in the public's mind, in fact was inspired by Jeanette Orrey, the catering manager of St. Peter's Primary School in Nottinghamshire. In 2000, Orrey, who was concerned about the health issues caused by poor quality industrially manufactured school meals, decided to replace them with in-house catering using local, organic food (Foley 2014, 201). Whilst it is true that the media tends to focus on the celebrity status of men, such as Oliver, overlooking or disregarding the inspiration and creativity of women—in this case, Jeanette Orrey—it is doubtful, given typical male propensity to appropriate women's ideas and creativity, that Oliver himself has given her credit for being the influence behind his campaign.

An increasingly common feature of television cooking shows featuring male celebrity chefs is that of promoting food-related issues, including food activism around issues such as healthy eating. Whilst, of course, in theory this is a good thing, it tends towards a moralising attitude towards women's domestic cooking and supposed failings in nurturing their families:

> In these shows the male protagonists use their high culinary capital to legitimate a disciplinarian attitude towards women, especially working-class mothers. The culinary shortcomings of these women are used to question their ability as mothers and 'to associate them with a range of moral and social failings'". (Leer 2017, 183)

The Male Takeover of the Domestic Kitchen

As men have become more involved in cooking, kitchens—traditionally a female domain—are becoming, especially in middle-class homes, increasingly claimed as a rightful place for men to express their culinary prowess. This, of course, has been facilitated by the open-plan design of many modern kitchens that are integrated into the living space, providing a platform for male 'foodie' cooks to show off their skills.

Men's increasing involvement in domestic cooking, albeit usually for special occasion cooking, rather than day-to-day food preparation, means that many kitchens are now being designed with men's needs and interests in mind, so that kitchens are increasingly becoming a more masculine space. For instance, men often insist on industrial-quality stoves and ovens that are more likely to be found in professional kitchens. In a study by British sociologists Angela Meah and Peter Jackson examining the increasing masculinisation of the domestic kitchen, one man they interview says that his new industrial-style French cooker "makes me feel really good about using it.…it's not a toy, it's a proper bit of kit" (Meah and Jackson 2013, 22). Even though some women too are drawn to expensive 'designer' ovens, as a status symbol, it is particularly men who are attracted to expensive gadgetry as a measure of their status and who are "particularly susceptible to appliance envy". As one man, writing in *The Age* newspaper, puts it: "There are just so many ways to tart up your kitchen with sleek gadgets and appliances. It's like kitchen porn – you fantasise about having the perfect kitchen" (Robinson 2002). In her gendered study of approaches to domestic cooking, Julie Parsons cites one of her male interviewees who "cooks only for special occasions", as having just bought 'a full width range/oven' because he says: "In doing up my first house in (country) there was only one concern – I wanted to have a say in the kitchen design – the cooker, etc., etc. – the rest of the house could be bright pink, I did not care as long as the kitchen looked good" (Parsons 2015, 42).

The increasing masculinisation of many domestic kitchens, both in the equipment used and in men's increasing presence, is resulting in what Meah and Jackson describe as a growing sense of women's alienation and marginalisation in what was previously a female domain. Women often feel intimidated and even humiliated by the determination of men to demonstrate their self-proclaimed high-level competence and expertise in the kitchen. The male appropriation of the domestic kitchen means that "women are not only engaged in a public struggle for equal power with men, but have also simultaneously lost influence in the private domain" (Meah and Jackson 2013, 21). Although it is true that some women, especially those that don't enjoy cooking themselves, welcome men's increasing role in cooking, the fact remains that men, in contrast to women, tend to take a very self-interested, individualistic, even competitive approach to domestic cooking. Unlike women, who are concerned with issues of family nutrition and preferences, men in general cook not because they have to but because they want to. One example that Meah and Jackson give from their interviews is of one 'foodie' man who enjoys cooking, but leaves the cooking of his fussy son's meals to his wife because for him "there is no incentive to make sausages and chips, it's simply not challenging enough" (20). This 'cherry-picking' by men and their approach to cooking as a hobby or leisure activity is corroborated by a number of other studies, including those mentioned earlier in this chapter.

Foodie Culture

In her analysis of foodie culture, Isabelle de Solier (2017) describes the motives driving the late twentieth and early twenty first century's phenomenon of 'foodie-ism'. Foodie-ism, in addition to being fostered by the rise of celebrity chef culture and food media, is in part a reaction against the industrialised food system. In post-industrial society, many people who are employed in non-manual employment, such as office work, feel alienated from the means of production, and therefore find meaning and satisfaction in activities that involve mind-body interaction such as cooking (de Solier 2017). Interestingly, whilst this renewed interest in cooking and other food-related activities, such as shopping at farmers markets, applies to both sexes, female foodies enjoy cooking because they feel less of an obligation to cook on a daily basis, as a result of alternatives, including eating out and ordering takeaways (de Solier 2017). This, of course, applies mainly to middle-class people who can afford to eat in restaurants on a regular basis.

De Solier found from her study of foodies that they viewed cooking, not only as a creative leisure activity, but also as a moral good and imperative; "all of my informants share a strong moral belief that people – especially foodies – should *enjoy* it – that is, a belief in the morality of cooking as productive leisure" (131, original italics). This attitude also extends to shopping which is viewed as an ethical issue as well as a source of pleasure, such as shopping at speciality food stores and farmers' markets. The development of farmers markets in recent years is, of course, a positive way of supporting small-scale organic farmers, as well as allowing urban consumers direct contact with rural food producers. Nevertheless, it does not in itself solve the problem of unequal access to healthy food; the foodie movement is essentially a very class-based, generally white phenomenon, with well-off middle-class consumers able to opt out of the industrially produced food system. As Julie Guthman lucidly points out in *Weighing In: Obesity, Food Justice and the Limits of Capitalism* (2011):

> The idea that the food system can be transformed by buying….good food (through informed choice) is a huge concession to the liberal idolatry of the market ….(and that) by changing what you eat may provide new and different pleasures – and I have little to quibble with that claim. But what to eat is the wrong question to ask in order to transform the food system into one that is more just and less toxic…. To the extent that you can buy your way out (of industrially produced foods), you are trying to purchase relative safety that others can't afford … (with) regulatory failure tending to accentuate class inequality rather than ameliorating it. (Guthman 2011, 148, 152)

Male 'foodies' are particularly influenced by the cult of male celebrity chefs, through their television programmes and cookbooks, both of which emphasise cooking as a masculine activity with little or no connection to the world of female domestic cooking. In an interesting examination of celebrity cookbooks and personas, Canadian sociologists, Josee Johnston, Alexandra Rodney and Phillipa Chong identify four types of hegemonic masculinities displayed by professional male food celebrities: the 'chef-artisan', exemplified by Pierre-Marco White, who present themselves and are presented by the media on a spectrum from artistic craftsmen to artistic geniuses; the 'gastrosexual', epitomised by Jamie Oliver, who embraces home cooking, while masculinising it with professional knowledge and expertise; the 'culinary maverick', who emphasises "an adventure discourse which effectively removes (himself) from both the home-kitchen as well as the standard culinary apprenticeship integral to the chef artists"; and the 'self-made man', who achieves celebrity status despite being usually from racially and class marginalised groups (Johnston et al. 2014, 18). Based on their analysis of celebrity chefs and their relationship to domestic cooking, Alexandra Rodney, Josee Johnston and Phillipa Chong, conclude that "discourses that masculinise men's cooking contribute to the perpetuation of gender norms, the maintenance of gendered power relations, and a hegemonic masculinity that privileges culinary professionalism, naturalises heterosexuality, and devalues the feminine credentials of domestic foodwork" (Rodney et al. 2017, 228).

A prominent feature of men's cooking vis-s-vis women's is the fact that when they cook for others they do it to impress, while women's cooking is very much bound up with caring. From their interviews with both male and female 'foodies', Cairns et al. found that "women tended to describe cooking for others through ideals of care that prioritise the daily nourishment of the family. By contrast, men's narratives framed cooking as a leisure activity, even when they were also engaged in daily food preparation at home" (Cairns et al. 2010, 605). They also found that male foodies seem to be particularly obsessed with gaining food knowledge and expertise, which they "often described as 'educating your palate'…,(with) their ever-growing food knowledge (being) a kind of cultural capital that positioned them apart from the average food consumer and served as a form of social distinction" (606). Even though women foodies were also interested in furthering their knowledge, it was men who emphasised the importance to themselves of knowledge and expertise, with male foodies being particularly impressed and influenced by professional male chefs (Cairns et al. 2010). They also found that both male and female foodies often referenced their mother's cooking as positive influences (2010); however, as I have elaborated elsewhere, in the case of male chefs and foodie home cooks, this rarely translates into respect for women's cooking as a whole.

Another distinguishing feature of male foodie cooking is the often single-minded focus on achieving the 'perfect' expression of a particular recipe. Cairns et al. cite one example of a male foodie's obsessive pursuit over six months, "trying to come up with the world's best lemon sorbet recipe" (606). Whilst this is undoubtedly an extreme example, it does illustrate male foodie's often obsessive approach to cooking. This stands in stark contrast to female domestic cooks who have neither the time nor inclination for such obsessive single-mindedness. Also, when men cook they are usually able to concentrate on the activity to the exclusion of all else; in contrast, women are constantly distracted by, and having to juggle, many other domestic demands on them, especially if they have children.

British academic, Julie Parsons makes an interesting distinction between 'foodieism' and 'epicureanism'; whereas both men and women can be defined as foodies, epicureanism is associated with the "performance of elite cultural capital and hegemonic masculinities" (Parsons 2015, 36). In a study she conducted on attitudes to food, shopping for food and cooking, based on 75 interviews with both men and women, she found that men distance themselves from the notion of feminised 'food work' by viewing it as 'food play', whereby "cooking is not considered 'work' at all, but represents a theatre for the performance and display of culinary expertise" (41). For most of her male interviewees, cooking was associated with the public, rather than the feminised 'domestic' sphere, as a source of adventure, risk taking and performing for an admiring audience. Furthermore, she found that the men in her study viewed food as "a social event and not a routinized everyday activity (and that they) were notably not concerned with health, dietary restraint or sustainability" (36).

It is sometimes argued/suggested that men's increasing involvement in domestic cooking is proof of increasing equality between the sexes with regard to housework. This, however, is refuted by the evidence presented in the studies cited in this chapter which confirm that when men do engage in domestic work, they generally do so very much on their own terms, choosing the more attractive aspects of domesticity, especially cooking, albeit usually on an occasional rather than daily basis. Indeed as Meah and Jackson (2013) found in their research, "there is little evidence ...of a significant transformation in gender roles and relations amounting to the 'democratisation' of domesticity" (24). In fact, they found that a frequent area of conflict within heterosexual relationships was the "additional mess that men's culinary activities produced", with the resulting "irony (being that) men's involvement in cooking...simply creates 'more work for mother'" (4; cited from Cowan 1983).

When men cook or start to cook—and especially when they are enthusiastic about it—they often behave as if they have discovered cooking themselves. This is because, unlike women, they are not part of a long line of culinary intergenerational continuity in the way that women are. Women know, even if only subconsciously or subliminally, that when they cook they are part of/carrying on an activity that innumerable generations of women before them have carried out over many generations/centuries/millennia. Because women know this on some level, when they cook it is not as if they have individually 'discovered' cooking in the way that men who cook often claim to have. This 'ownership' by men of cooking comes from both ignorance and male arrogance, notwithstanding some notable exceptions (for instance, Antonio Carluccio) who acknowledge the importance of the female domestic culinary tradition, and the influence of their own mothers and grandmothers. When men 'discover' cooking it does not come loaded with the 'baggage', both negative and positive, and the complex history and often contradictory messages that women inherit. For men, cooking is not associated with a history of serving/servicing others as it is for women, who sometimes do this willingly and happily and sometimes reluctantly and resentfully—but in either case rarely have a choice. Being free of these negative connotations, men feel free to claim for themselves the creative and enjoyable aspects of cooking. This male tendency to 'cherry-pick' is deeply problematic. It is fundamentally dishonest to claim to be egalitarian when they choose to do the cooking, usually on an occasional basis, and not the less pleasant jobs of cleaning, which are invariably done by women, either the female partner or spouse or hired female help.

This is not to suggest that men shouldn't cook or know how to cook; after all, knowing how to cook is an essential life-skill, and men should be able to do this for themselves and for others so that women are not expected to service them in this way. However, when men cook they should acknowledge and respect the *long intergenerational female tradition of domestic cooking*, a tradition that has essentially contributed to/created all the innumerable regional cuisines of the world that have evolved out of a particular climate, environment, social, cultural and historical context. This is one reason why some men's claim to be innately superior cooks to women, despite the occasional nature of their cooking, is so absurd. Such a claim neither recognises nor values the body of accumulated food and cooking knowledge and the multifaceted skills involved in preparing family meals on a daily basis. Indeed, denial and/or trivialisation of women's culinary skills and knowledge may actually mask a resentment or envy on the part of some men at these fundamentally female skills and knowledge.

It may be thought that men's increased involvement in domestic cooking is evidence of an increase in equality with regard to the sexual division of domestic labour. However, as I have demonstrated in this chapter, this is not the case, and in fact, on the contrary, represents a reinforcement of inequality, with men appropriating from women the one area, cooking, that has the potential to be deeply creative and satisfying. This has become even more attractive to some men since cooking has become a trendy middle-class activity promoted by lifestyle television programmes and cookbooks authored by celebrity male chefs.

During the time of the COVID-19 pandemic, it is very interesting to note that during the forced lockdown of this period, more men have taken up cooking as a creative recreation activity. Of even more interest/significance is that this appears to be a cross-cultural phenomenon with published articles describing the trend not only from Western countries, but also from countries such as India and Saudi Arabia, places where, traditionally, men rarely set foot in domestic kitchens. One article from the Hindustani Times, 'Quarantine makes cooking cool for men' describes how many Indian men, encouraged by lockdown spiritual leader Sri Sri Ravi Shankar's blog # let men cook, have turned to cooking. One enthusiast, in true masculine fashion, posts: "I'll come out as next Master Chef of India after this lockdown. I bet! It has become more (of a) kind of game every day to cook something new and more delicious" (Sehgal 2020). A Saudi article reports that "the famed cooking skills of Saudi women have found a surprise challenger during the COVID-19 lockdown – with men revealing their hidden culinary talents." Again, bringing a masculinist approach to cooking as a competitive activity, one young Saudi man who intends to continue cooking post lockdown, has introduced a cooking competition challenge for his family and cousins. However, even when men's cooking isn't viewed as a display and competition, they congratulate themselves on managing the trials of cooking for children. One article in the British newspaper, *The Observer* is by a man who has discovered the frustrations of feeding two young children on a daily basis during lockdown. He says that he suspects that the greatest challenges (apart from for those on the frontline and those with the virus) of the past few weeks, and "moments of profound despair, have come in the kitchen" (Tim Lewis, *The Guardian*, 17 May 2020). It is telling that nowhere does he mention that this challenge of making sure their children are properly fed is one that women throughout the world do on a daily basis, year in year out. Cooking for most women is not optional; however, it is precisely women's daily domestic cooking that has built up, over many generations, the basis of the world's innumerable food cultures. In the following chapter, I delve into the largely unexplored world of female intergenerational transfer of food knowledge and skills.

References

Ashley, Bob, Joanne Hollows, Steve Jones, and Ben Taylor. 2004. *Food and Cultural Studies*. London and New York: Routledge.

BBC News. 2011. "Hay Festival: Nigella Lawson's Husband Prefers Weetbix." BBC News, 29 May.

Besen-Cassino, Yasemin, and Dan Cassino. 2014. "Division of House Chores and the Curious Case of Cooking: The Effects of Earning Inequality on Household Chores among Dual-Earner Couples." *International Journal of Gender Studies* 3 (6): 25–53.

Blythman, Joanna. 2006. *Bad Food Britain: How a Nation Ruined Its Appetite*. London: Fourth Estate.

Cairns, Kate, Josee Johnston, and Shyon Baumann. 2010. "Doing Gender in the Foodie Kitchen." *Gender and Society* 24 (5): 1480–1523.

Charles, Nickie. 1995. "Food and Family Ideology." In *The Politics of Domestic Consumption: Critical Readings*, edited by J. Stevie Jackson and Shaun Moores, 100–115. London: Prentice-Hall; New York: Harvester Wheatsheaf.

Charles, Nickie, and Marion Kerr. 1988. *Women, Food and Families*. Manchester and New York: Manchester University Press.

COVID -19 Lockdown Brings Out Hidden Kitchen Talents of Saudi Men. 2020. Arab News, 10 April. https://arab.news/c28bm.

Cowan, Ruth Schwartz. 1983. *More Work for Mother: The Ironies of Household Technology from the Open Hearth to the Microwave*. New York: Basic Books Inc.

De Solier, Isabelle. 2017. *Food and the Self: Consumption, Production and Material Culture*. London and New York: Bloomsbury Academic.

DeVault, Marjorie L. 1991. *Feeding the Family: The Social Organization of Caring as Gendered Work*. Chicago and London: The University of Chicago Press.

Ellis, R. 1983. "The Way to a Man's Heart: Food in the Violent Home." In *The Sociology of Food and Eating: Essays on the Sociological Significance of Food*, edited by Anne Murcott, 164–171. Aldershot, Hants: Gower.

Foley, Caroline. 2014. *Of Cabbages and Kings: The History of Allotments*. London: Frances Lincoln Ltd.

Frye, Marilyn. 1983. *The Politics of Reality: Essays in Feminist Theory*. Trumansburg, NY: The Crossing Press.

Guthman, Julie. 2011. *Weighing In: Obesity, Food Justice, and the Limits of Capitalism*. Berkeley and Los Angeles: University of California Press.

Hollows, Joanne. 2002. "The Bachelor Dinner: Masculinity, Class and Cooking in Playboy, 1953–1961." *Continuum: Journal of Media and Cultural Studies* 16 (2): 143–155.

Hollows, Joanne. 2003. "Oliver's Twist: Leisure, Labour and Domestic Masculinity in *The Naked Chef*." *International Journal of Cultural Studies* 6 (2): 229–248.

Innes, Sherrie A. 2001. *Dinner Roles: American Women and Culinary Culture*. Iowa City: University of Iowa Press.

Johnston, Josee, Alexandra Rodney, and Phillipa Chong. 2014. Making Change in the Kitchen? A Study of Celebrity Cookbooks, Culinary Personas, and Inequality. *Poetics*. http://dx.doi.org/10.1016/j.poetic.2014.10.001.

Kemmer, D. 1999. "Food Preparation and the Domestic Division of Labour Among Newly Married and Cohabiting Couples." *British Food Journal* 101 (8): 570–579.

Leer, Jonatan. 2017. "If You Want to, You Can Do It! Home cooking and Masculinity Makeover in *Le Chef Contra-Attaque*." In *Food, Masculinities and Home: Interdisciplinary Perspective*, edited by Michelle Szabo and Shelly Koch, 82–196. London: Bloomsbury Academic.

Lewis, Tim. 2020. "Family Cooking Under Coronavirus: 'I've Become a Chef for Two Grumpy Regular Customers'." *The Observer Newspaper*, 17 May.

McFeely, Mary Drake. 2000. *Can She Bake a Cherry Pie: American Women and the Kitchen in the Twentieth Century*. Amherst: University of Massachusetts Press.

Meah, Angela, and Peter Jackson. 2013. "Crowded Kitchens: The Democratisation of Domesticity." *Gender, Place and Culture*. https://doi.org/10.1080/0966369x.2012.701202.

Murcott, Anne. 1995. "'It's a Pleasure to Cook for Him': Food, Mealtimes and Gender in Some South Wales Households." In *The Politics of Domestic Consumption: Critical Readings*, edited by Stevi Jackson and Shaun Moores, 89–99. London and New York: Prentice-Hall/Harvester Wheatsheaf.

Neuhaus, Jessamyn. 2003. *Manly Meals and Mom's Home Cooking: Cookbooks and Gender in Modern America*. Baltimore and London: The John Hopkins University Press.

Parsons, Julie, M. 2015. "The Joy of Food Play—An Exploration of the Continued Intersectionalities of Gender and Class in Men's Auto/Biographical Accounts of Everyday Foodways." Special Issue of *Women, Gender and Research* (3–4): 35–47.

Robinson, F. 2002. "Sexing the Kitchen." *The Sunday Age*. Melbourne, Australia.

Rodney, Alexandra, Josee Johnston, and Phillipa Chong. 2017. "Chefs at Home? Masculinities on Offer in Celebrity Chef Cookbooks." In *Food, Masculinities and Home: Interdisciplinary Perspectives*, edited by Michelle Szabo and Shelly Koch, 213–230. London: Bloomsbury Academic.

Sehgal, Kritika. 2020. Quarantine Makes Cooking Cool for Men. *Hindustani Times*, 14 April.

Shapiro, Laura. 2005. *Something from the Oven: Reinventing Dinner in 1950s America*. New York: Penguin Books.

Short, Frances. 2006. *Kitchen Secrets: The Meaning of Cooking in Everyday Life*. New York and London: Berg.

Van Berkel, Michel, and N. D. De Graaf. 1999. "By Virtue of Pleasantness: Housework and the Effects of Education Revisited." *Sociology* 33 (4): 785–808.

CHAPTER 5

Cultural Memory and Female Intergenerational Culinary Culture

In an essay on cooking as a practical everyday expression of culture, French sociologist Luce Giard describes how as a child she deliberately refused to learn how to cook because she associated it with female subordination. However, as a young adult the experience of eating dull and tasteless institutional food whilst at university made her appreciate the care and quality of her mother's cooking, and regret that she had refused to learn from her. As a result, she now realised the importance and potential pleasure of cooking. Whereas previously she had regarded as "elementary, conventional, and pedestrian (and therefore a bit stupid) the feminine *savoir faire* that presided over buying food, preparing it, and organising meals", she now discovered that cooking is not simply a question of carrying out routine tasks, but also involves many skills, plus intelligence, imagination, planning, creativity and innovation. Setting out to teach herself how to cook from a book, she found to her surprise that, despite her refusal to learn from her mother, she had absorbed, as though by osmosis, a "sense memory [that] had kept track of certain tastes, smells, and colors", an embodied memory of the sound of bread being kneaded, foods frying and simmering. She discovered that she was "the heiress and guardian without wanting to be…*[of] a woman's knowledge…that integrated me into the great corps of women of my lineage, incorporating me into their anonymous ranks*" (Giard 1998, 153; italics added).

V. A. Swinbank, *Women's Food Matters*,
https://doi.org/10.1007/978-3-030-70396-7_5

In this chapter I describe how, despite women's sometimes conflicted relationship with the daily responsibility of having to prepare family meals, this has, over countless generations, created the basis of the vast culinary repertoire that exists throughout the world.

'A Thread of Continuous Knowledge': The Intergenerational Transmission of Female Culinary and Food Knowledge

Traditionally, knowledge of food and cooking has been handed down from grandmother and mother to daughter and granddaughter in a long continuous line linking the generations. In this way culinary knowledge has been accumulated and transmitted over many centuries, contributing to the development of the countless regional cuisines throughout the world. As Carol Field puts it, describing the development of the distinctive regional cuisines of Italy: "Women took whatever was growing in the kitchen garden and the fields, stored in the larder or plucked straight from nature, and *with imagination, frugality and a deep-seated knowledge of local ingredients created an entire cuisine*" (Field 1997, 14; italics added). Given the importance of food to human culture and indeed to human survival, this largely female intergenerational food-related knowledge and expertise constitutes a major contribution to civilisation. The transmission of this knowledge and these skills has historically been principally oral and also by girls observing their older female kin-mothers, grandmothers and aunts and great aunts.

This culinary apprenticeship, described by David Sutton, an anthropologist studying the transmission of food knowledge on the Greek island of Kalymnos, is an 'embodied apprenticeship' (Sutton 2001, 126) based on actual doing, a mind–body relationship, referred to by Giard as 'doing cooking', a way of learning which stands in contrast to learning from texts or cookbooks. The knowledge and know-how involved in food preparation and provision is accumulated over many years of actually doing the many activities both mental and physical that are involved in this. As Sutton explains from his study of the transmission of culinary knowledge on Kalymnos, an embodied knowledge such as cooking or dancing is learnt by watching others who are skilled in the particular activity and then doing it oneself. The embodied learning involved in learning to cook has been referred to as 'the habit of hands', whereby the routines and regular practice of food preparation become an embodied memory. Such a process is illustrated in the story of a Greek grandmother whose "recipes are written into her hands, into the strata of her calluses", so much so that, when she becomes blind in old age, unable to prepare all the dishes she once made, and mourning "the days when the kitchen was hers", nevertheless, "even with her sight gone completely, occasionally a dish miraculously appears from the kitchen – *pastitsio* or *domates me agva* prepared by hands that remember" (Kapsalis 1997, 32).

Lisa Heldke describes cooking or foodmaking as a 'thoughtful practice', representing a synthesis of the mental and physical or manual which consists of "a 'mentally manual' activity [and] a 'theoretically practical' activity. This 'bodily knowledge' is acquired through 'embodied experience' and comes from "seeing ourselves as connected to the things we grow and cook…[and] is an acknowledgement of the fact that I *know* things literally with my body, that I 'as' my hands, know when the bread dough is sufficiently kneaded…[T]he knowing is 'contained' not simply 'in my head' but in my hands, my wrists, my eyes and nose as well" (Heldke 1992, 218). The social nature of cooking in societies such as Kalymnos, where women often share certain food preparation tasks, enables the development of a body of food and culinary knowledge to develop, with women discussing and comparing recipes, ingredients and ways of doing things. This food talk is part of women's everyday conversation, building up a stock of food and cooking knowledge that comes to constitute a locally distinctive regional cuisine. This is transmitted between women of the same generation and from one generation to the next, in an "apprenticeship [that] becomes a site of transmission in the broader sense of a woman's culture, history and everyday experience" (Sutton 2001, 137).

In her classic book on Middle Eastern cooking and food culture, Claudia Roden describes how: "Mothers and daughters, sisters, cousins and friends love to talk about what they will serve their family for lunch or dinner, and they sit or help each other to prepare delicacies which require time and skill" (Roden 1986, 47). Far from being a trivial matter, this is all part of how food cultures are created and preserved. The social nature of a cuisine is dependent upon the existence of continuity and community, requiring "a thread of continuous knowledge [that] underpins standards of traditional production, preparation and consumption…[and which] are passed on from one generation to the next" (Mars 2000, 157). Heldke points out that creating, exchanging and testing recipes are community activities (Heldke 1992), with the handing on of food knowledge from one generation to the next both depending upon community and helping to create and perpetuate it. In traditional societies, such as those described by Sutton and Roden, this sharing of knowledge often takes place in shared or public spaces such as courtyards where women share the tasks of sorting through beans, peeling vegetables or time-consuming activities which are made easier by sharing, such as rolling *dolmades*, whilst they talk and share information. No doubt this talk is often dismissed by masculinist thinking as 'trivial women's gossip', but this 'gossip' and information sharing makes a vital contribution to the body of food knowledge that develops into a regional cuisine. Indeed, so important is the communal and social nature of food preparation in many traditional societies, that women sometimes resist the tendency to more private, isolated kitchens in modern houses. An example of this is of new housing developments in Sicily in which the kitchen is tucked away at the back of the house, preventing the socialising that traditionally takes place between women as they prepare food. In order to overcome this, the women converted their garages into kitchens—against building regulations—so that they could "restore the familiar flow of neighbourhood social interaction that once occupied collective courtyards" (Birdwell-Pheasant and Lawrence-Zuniga 1999, 23). In a similar fashion, Sutton describes how women on Kalymnos prefer to sit outside whilst they prepare food, such as cutting vegetables, with a bowl on their laps, skilfully turning the vegetables as they are cut. This not only helps avoid back pain caused by standing at a bench, but, also importantly, enables them to socialise with neighbours and friends, and "even mak(ing) processing food an occasion for sitting in a circle and sharing news and stories" (Sutton 2014, 53).

The social aspect of food preparation and cooking described above typifies the approach to food that is a central feature of societies such as Kalymnos where "cooking is a shared value" and that cooking "in and of itself matters"; it is therefore considered "an everyday significant practice that generates…much discourse *precisely because it matters*" (182; italics added). It is this kind of social exchange and discussion that forms the basis of a vibrant food culture, and as Sutton points out, citing Sidney Mintz, people's shared investment in talking about food and cooking is an integral part of creating and maintaining a distinct regional cuisine, such as that on Kalymnos. Both men and women have a vested interest in the food culture of Kalymnos, with both sexes having strong opinions on food and cooking, although generally men as eaters and women as cooks.

As elsewhere in the world, it is women who are the principal preparers and cooks of domestic meals, with men doing occasional cooking for 'special occasions' or for themselves when their wives are absent. In addition, a common theme, as elsewhere (which I have analysed and critiqued in this book), is Kalymniot men's insistence that they are better cooks than women, with one man who cooks as a hobby, claiming that "men on Kalymnos tend to be better cooks than women because 'for women, cooking is a routine, so they get tired of it, and they lose the feeling that they are doing something special. Whereas for men, since they don't cook constantly, they are more interested in what they are doing'" (2014, 180).

'More Precious Than Jewels': Cookbooks in the Lives of Women

Embodied food knowledge is transmitted primarily orally and by observation and participation, in contrast to learning from written recipes and cookbooks. Nevertheless, for literate women the writing of a personal cookbook has long been an important means of not only recording recipes, but also often as a kind of personal diary and an important way of expressing their thoughts and opinions as well as their culinary creativity. As Janet Theophano puts it:

> There is much to be learned from reading a cookbook besides how to prepare food – discovering the stories told in the spaces between the recipe or within the recipes themselves…..The themes found in cookbooks are timeless…[they] have thus served women as meditations, memoirs, dairies, journals, scrapbooks and guides. (Theophano 2002, 6)

The study of women's cookbooks over the centuries not only informs us of recipes of the past, but is also an extremely valuable way of gaining some understanding and insight into the lives of women in the past, especially as historically women are largely invisible.

Cookbooks are also an extremely important means of preserving recipes and food knowledge, particularly as modernisation and industrialisation of food threatens the survival of traditional food systems and their associated cuisines. When Rosemary Brissenden came to rewrite her book *South East Asian Food* almost thirty years after it was originally published in 1965, she found that, whilst some of the changes she observed in cooking practices were inevitable with modernisation and not necessarily negative, for instance certain shortcuts and techniques, she was nevertheless alarmed at the "loss of domestic heritage, loss of time to resuscitate it, loss of skill and culinary control" (2007, 2). Realising that domestic culinary traditions were becoming endangered as a result of lack of time in the "hurly-burly of middle-class, urbanised daily life", and the increasing influence of western fast foods such as hot dogs, hamburgers and pizzas, Brissenden was anxious to include in her book traditional dishes "that could be lost before too long" (30). Professor of African history, James C. McCann, in his book on the history of African cuisine, describes how girls traditionally learn "the art of cooking by listening and watching mothers, aunts, relatives and neighbours….(and that) the transcribing of that knowledge as a formal written recipe….violates, in a sense, their original flexible and creative oral character". However, for historians written recipes are "valuable as markers of accumulated experience of those generations of cooks who have gone before" (McCann 2010, 11). Nevertheless, a healthy and vibrant food culture cannot be based solely on a body of written knowledge preserved in a museum-like fashion, but needs to be practised. Cuisine is sometimes compared to language with the same day-to-day practice and face-to-face interaction between people necessary to its continuation and survival. Indeed, "[K]nowledge of cuisine and its structure is like knowledge of a language; it extends beyond mere individuals and resides within a people's culture…[and] though cultures are the repositories of knowledge about cuisine, it is in face-to-face groups within a culture – in its villages, neighbourhoods and families – that cuisines are fostered" (Mars 2000, 157).

Learning to cook from books requires at least a basic practical hands-on knowledge of cooking which has been learnt, or at least absorbed on some level, by osmosis, by a sensory memory, such as that described by Giard—that is, by being around cooking as an actual activity, not as something out of a tin, or a ready-made processed frozen meal 'cooked' in a microwave. While written recipes can act as an 'inspiration' or 'memory jogger', to learn to cook well one needs a 'hands-on' teacher, because "[W]ords can't explain what you must learn using your hands and nose and mouth" (Heldke 1992, 219). Even in societies with a strong and robust culinary culture where women still cook instinctively—having a 'feel' for cooking born of constant practice—and scorn the notion of depending on written recipes, there is a long tradition (at least in literate societies) of notebooks of handwritten recipes passed down as treasured family heirlooms from one generation of women to another. Apart from the importance to individual family members of such a record, these culinary notebooks are also an invaluable source of ethnographic culinary and food knowledge, more important than ever in the face of rapidly changing food habits due to the processes of industrialisation, urbanisation and globalisation.

In her study of the transmission of cooking knowledge involving interviewing forty-eight migrant women in Western Australia, Sian Supski found that they had "learned to cook from mothers, mothers-in-law, grandmothers, sisters, servants, aunts, female friends and neighbours, and occasionally fathers, fathers-in-law and husbands...[with] recipes ...learnt by heart, written down and reproduced through generations from mother to daughter (or grandmother to granddaughter)" (Supski 2005, 86). A liking of cooking was often passed from mother to daughter, with the majority of these women cooking their everyday meals from memory, being confident enough to do so without needing written recipes. Nevertheless, cookbooks were important to them for "speciality cooking, such as cakes, pastries, desserts and biscuits, where recipes were used to ensure correct measurements" (86). Cookbooks were also important both as a source of pleasure, many women collecting them as a hobby, and as a means of adding to their existing knowledge, as a source of inspiration, as well as maintaining a connection to their female forebears and their cultural heritage. For some women, the creation of their own recipe books was "a means of connecting...with other family members and a link between generations" (90). One graphic example of this is a memoir written by an Italian-Scottish woman, addressed to her daughter in order to record and preserve her culinary and cultural heritage handed down from her grandparents who emigrated to Scotland from southern Italy. Stating that she has "no material things left from them to give to you", she continues: "[S]o dear Francesca, here in these pages is that heritage of your family that was handed down to me by word of mouth, with humour and love. Treasure it; it's more precious than jewels" (Contini 2002, 9). According to culinary historian, Janet Theophano, traditionally: ["T]he offering of a recipe book was a sign of affection, a tie binding together women of different generations" (Theophano 2002, 86).

From the migrant women Supski interviewed, she that "for all [of them] maintaining family recipes ...is highly emotional" (Supski 2005, 91). A particularly poignant example of this is of an Italian Jewish woman who, on emigrating to Australia, was given, as a 'loving gesture' the much valued personal cookbook of her mother-in-law, one of only two family members to survive the Holocaust, and who "used to read recipes from her cookbook in order to compensate for starvation during the war" (91). An even more poignant example is that of the women of Terezin (renamed Thereisenstadt by the Germans), a Jewish ghetto created by the Nazis near Prague. The women inmates of this ghetto, of whom very few survived, attempted to deal with their hunger and terrible suffering by compiling a cookbook in a "spiritual revolt against the harshness of …conditions…[and] a flight of the imagination back to an earlier time when food was available, when women had homes and kitchens and could provide a meal for their children" (De Silva1996, xv–xvi). In writing this cookbook, the women inmates of Terezin were also attempting to preserve a tradition, common in Europe at the time, of passing on personally handwritten cookbooks to their daughters. Mina Pachter, who had been involved in writing it and who did not survive the harshness of the ghetto, entrusted the manuscript to a friend to deliver it to her daughter who had escaped the Holocaust by emigrating to Palestine. When the manuscript eventually found its way to her daughter twenty-five years later and who was by then living in the United States, she felt as though: "After all those years, it was like her hand was reaching out to me from long ago" (1996, xxvii).

In her book *In Nonna's Kitchen*, Carol Field (1997) says that of the large number of women she spoke to from a wide variety of Italian regions and backgrounds, ranging from poor peasants to well-off middle-class women, all of them "shared a passion for food…[and] a devotion to the food of their families, of their villages, towns and cities….[T]he food the women cook, the dishes they learned from mothers and grandmothers and aunts, is a profound link to their pasts,…inflected with regional and social nuances" (6–7). Field describes the high value that is placed by women on kitchen notebooks handed down to them from their grandmothers; these are kept "carefully in a separate drawer…for they are the source of family dishes that pass from generation to generation" (11). The importance of this female intergenerational food knowledge is expressed by Australian cookery writer, Margaret Fulton, in her acknowledgement of the culinary legacy of her Scottish mother:

> My passion for cooking was fuelled, first and foremost, by the sheer joy of eating all the sublime things that came out of my mother's kitchen…What gives me enormous happiness is to see the love of food and passion for cooking passed on from my mother and through me to my daughter… (Fulton 1998, 7)

The emotional significance of intergenerational cooking knowledge to women—and its lack of recognition by many/most men—is illustrated by the story cited by Supski of two sisters who mourn the loss of their mother's thick exercise book with its handwritten recipes thrown out by their father after the death of their mother. They mourn the loss of certain recipes that they associate with their mother's special way of cooking them, for instance, a particular fruit-cake recipe which they now will never be able to reproduce (Supski 2005, 90). Women often also feel a connection to previous generations through the use of a particular pan, bowl, rolling pin or other kitchen utensils that belonged to their mother or grandmother. As one woman expresses it: "When I want to talk to my grandmother...I use her implements and I cook her dishes" (Field 1997, 3).

The writing of a cookbook/compiling of recipes can in itself generate a sense of community and sharing of recipes and culinary knowledge—this is especially the case with women's cookbooks which often draw on the shared knowledge of other female cooks—mothers, grandmothers, aunts, friends and neighbours. One such example is that of Palestinian cookbook author, Joudie Kalla (2018), who describes how when she talks in depth to her mother about recipes:

> She starts talking about another and another that spin off from the first. The number of recipes stored in her head is simply unbelievable....The bond between cooking and memories, creating flashbacks to another time, is something I am so pleased to have been a part of with her and my twenty aunties, who are all so excited to be resurrecting recipes they too had forgotten. (10)

Women's cookbooks are often collective or collaborative endeavours. In her study of women's cookbooks, Theophano (2002) found that personal notebooks of recipes handed down through the female line often contained the recipes and notes of several women, with friends or female relatives adding their own recipes or variations on existing ones to it. Often such cooking notebooks consist of a kind of scrapbook with recipes written down by different women stuck into it: "[O]ver the centuries, in the domestic sphere, the activities related to food have been primarily in the woman's domain. In the course of day-to-day life, by exchanging recipes and other household advice, women generated their culinary knowledge collaboratively and wrote their cookbooks co-operatively" (13). Giving the example of a letter from Jane Austen to a friend requesting a recipe that she had enjoyed during a visit, Theophano points out that over the centuries literate women have frequently written to friends requesting the details of a particular recipe. Interestingly, the word 'recipe' (or 'receipt' in earlier times) is derived from the Latin *recipere*, meaning 'to exchange'. As such, the exchange and sharing of recipes between women is an act of trust and friendship (whilst conversely the withholding of such indicates the opposite), a "way to signal and affirm affiliations...cement[ing] relationships fostered in daily life" (41).

The transmission of written recipes, as well as household and medicinal knowledge and remedies has a long history. Starting with the Elizabethan Age, the handing on of a personal cookbook amongst the literate upper classes was not only for sentimental reasons, but was also considered essential to a young woman's education in the art of cooking, as well as in other vital household matters of the time, including medicinal remedies and the art of distilling. As historian of cookbooks, Janet Theophano (2002) explains: "Women across the social spectrum were engaged in healing activities...[with S]ome middle-ranking women earn[ing] their livelihoods as midwives and healers" and that "All chatelaines of great houses were expected to be conversant with medicine, the arts of distilling cordials and healing waters and the cooking of restorative broths" (93). Significantly, the early origins of modern chemistry probably lie in what was called 'kitchen physick', with "many noblewomen engaged in scientific research and medical experimentation while occupied in domestic activities". Indeed, it is thought that the famous seventeenth-century chemist, Robert Boyle, probably obtained many of his ideas from the recipes of his sister, Lady Ranelagh (93).

'Ties to Homelands Left or Lost': Food and Migrant Identity

The importance of food and cooking in maintaining a sense of cultural identity and connection to one's homeland is of great importance to migrant communities. Indeed, "[M]igrants the world over bring their dietary preferences and cooking practices with them....Food is vested with symbolic ties to homelands left or lost" (Carney and Rosomoff 2009, 185). This longing is expressed by Abla Amad (2001), who, upon her arrival as a young migrant to Australia in the 1950s, assuaged her homesickness by cooking Lebanese food taught to her by other women of Melbourne's Lebanese community, who she refers to as her 'aunties'. Italian migrants, who along with many other nationalities, settled in Australia in the post-Second World War years brought with them those aspects of their culture, most notably food, that would enable them to 'feel at home' in an alien environment, growing vegetables and fruit trees in their urban gardens. Over time their food culture greatly enriched Australian eating habits (Swinbank 1995). In a cookbook recording recipes and food traditions of their Calabrian background/heritage, Teresa Oates and Angela Villella (2011) describe how their families, who migrated to Australia in the 1950s and 1960s, brought with them their rich culinary history; their book "celebrates the one constant of our families' journeys – their food (which carries) a sense of who we are and where we came from" (viii).

Supski (2005) found that for migrant women cooking their traditional foods was especially important in helping them cope with the stress of adapting to a new country, and that the "passing-down of recipes is an attempt by women to ensure that cultural traditions are not lost" (91). Nevertheless, even though they may wish to record their recipes, some migrant women do not want to insist on their children continuing their traditions because of their concerns about the assimilation of their first-generation children into the culture of the adopted country. However, whilst an interest in cooking traditions might skip a generation, fortunately "an interest in the traditional recipes of the family,...was often acquired by grandchildren, granddaughters in particular" (92).

Their traditional food is important not only to migrant and refugee women's ethnic and cultural identities, but also provides a way of feeling more involved in the life of their adopted country, as well as sometimes as a means of survival. For example, Lorraine McGinniss describes how a group of refugee women in Western Australia, whose places of origin include Afghanistan, Iraq, Vietnam, as well as African and Latin American countries, meet every week to share with each other the cooking and stories of their homelands, and how despite their cultural and language differences, food and cooking provide a common bond, with the women eager to learn about the different cuisines from each other. This acts as an important means of social support as well as connecting with other women, by providing the "camaraderie and the company the women enjoy as they cook [which] are often the things they miss most about their faraway homelands" (McGinniss 2005, 5). McGinniss explains how all of these women, who have learnt cooking from their mothers and other female relatives, "have an extraordinary depth of knowledge of food and a seemingly endless repertoire of recipes which they carry with them in their heads" (5).

Describing the centrality of women's shared food preparation—especially in the labour-intensive task of making tamales—in a Mexican migrant community, anthropologist Brett Williams explains that for these migrant workers food is not only nurturance but also an extremely important way of maintaining identity and culture in an alien environment, as well as a means of negotiating relationships both inside and outside the community—and in all of this women are the principal players: "Migrants use foodways to preserve a sense of who they are in an alien cultural setting just as they mobilize foodways to approach and appraise friends, and ….it seems that women purposefully monopolize those skills necessary for plying and obligating others and for keeping ethnicity alive" (Williams 1984, 122).

Although these women assert their monopoly of domesticity and food preparation, ultimately they are subordinate to their husbands whose authority they do not question. Williams points out that even though this embrace of domesticity would appear to confirm their subordination, nevertheless the making of food, especially the ritual and cooperation involved in making tamales, is in fact a survival strategy for these migrant workers, a way of supporting extended families and holding communities together, with women being central to this:

> Just as tamales underscore women's domestic commitments, the everyday preparation and sharing of food routinely reaffirms family ties and allows families to work as efficiently and profitably as possible. Especially in emergencies, the sharing of food attests to migrants' visions of their lives as closely, mutually intertwined…offer[ing] …creative domestic cooperation in stretching and sharing food within families and in the continuing migration of family members north from Mexico and back and forth to Illinois to work. (117)

These migrant women also use food to negotiate relations with the outside world, which Williams points out may be crucial to the maintenance of family life in an alien environment. For instance, in needing to negotiate with and befriend Anglo-Americans who have the resources and power, these migrant women may ply them with tacos and tamales, an acceptance of which indicates "a show of respect for Tejano culture and as a tentative commitment to kin-like relations" (122). Therefore, for these migrant women food is not only a means to cement and strengthen relationships with kin, but also an important way of exercising influence in the world outside the home. As such, it is "misleading... to proclaim family life as an isolating, stultifying, belittling activity for women, as [these] women use food to make friends and allies as well as to identify outsiders who will or will not commit themselves to the Tejano family concerns" (122). This example highlights just how complex an issue food and cooking is for women—particularly for women in heterosexual relationships—and their often contradictory relationship to it; how, on the one hand it can be a source of influence and authority, and on the other of servitude and subservience to men.

'Dishes that Remind Them of Home': Refugees and Culinary Cultural Memory

The displacement and dispersal of populations caused by war have major negative repercussions in most aspects of people's lives, not least of which is access to sufficient and culturally appropriate food. This represents a serious threat to the maintenance of regional culinary traditions that have been developed, mostly by women, over many generations. Nevertheless, when all else is lost—homes, livelihoods, not to mention lives—attachment to their particular food culture becomes especially important to displaced people, as food memories are the one thing that can't be eradicated, remaining intractable to the preservation of people's identity and culture. And it is invariably women, in their role as cooks, preparers and preservers of food, who are the repositories and transmitters of this vital food and culinary knowledge. Unfortunately, precisely because this is primarily female knowledge and expertise, it tends to remain largely unacknowledged and to 'fly under the radar'. However, there are now a number of female-authored cookbooks which address this important issue.

One such striking example is that in Caroline Eden's *Black Sea* (2018) in which she describes how the loss of culinary knowledge resulted from ethnic cleaning by the Russians in the North Ossetia region of the Caucasus in the nineteenth century; but how the ancient traditional recipes, which are now virtually extinct in North Ossetia have been preserved and maintained by the female descendants of the displaced women who now live along the Turkish Black Sea coast (Eden 2018, 126). Another recent example is that of refugees from war-torn Syria who have been dispersed across the world, but who are determined to maintain their rich cultural heritage. Two young London-based Syrian women have produced a book in which they interview displaced Syrian women who, no matter where they are, and even in the most trying circumstances, are determined to maintain their culinary heritage. They state that:

> [O]ur culture is under attack as never before....and our people are fleeing their homeland, disappearing around the globe to seek a new, safer life. This is possibly the country's darkest hour, but even now in tiny flats in Beirut, Berlin and Bradford, Syrian families are searching out the best tomatoes and lemons, pomegranates and parsley to recreate the dishes that remind them of home. (Itab Azzam and Dina Mousawi 2017, 7)

The great expert on Middle Eastern and Jewish food cultures, Claudia Roden (1986), recounts how, after her family and relatives were expelled from Egypt in 1956, whenever they got together in their new homeland of England or France, one of the first things that women would want to discuss were treasured family recipes. She describes how, as a schoolgirl in Paris, she and her brothers would go to a relative's flat every Sunday to eat the Egyptian dish *ful medames*, and how this simple peasant meal of fava beans with olive oil, lemon juice and a boiled egg "became invested, with all the glories and warmth of Cairo, our home town, and the embodiment of all that for which we were homesick" (13). She saw her job as examining women's home cooking, and was inspired to start recording the recipes before they were lost.

Perhaps one of the most striking examples of women's struggles and ingenuity in trying to feed their families and preserve their food culture is that of the women of the Gaza Strip, who live in exile in their own land, and are virtually imprisoned within a tiny piece of territory which is constantly under siege and attack. Due to the bombing of essential infrastructure by Israeli forces and the repeated closure of borders by both Israel and Egypt, about 80% of the Gaza population is food insecure (El-Hadad 2012, 56). Despite the terrible privations, with the UN providing rations of flour, beans, sugar, salt and powdered milk—in contrast to the wealth of fresh produce such as vegetables, fruit and fish prior to the Israeli occupation—Palestinian women "have adapted to this reality, creating innovative dishes with what ingredients are available" (61). This adaptation includes making yoghurt and cheese from powdered milk rations, and the substitution of the once plentiful olive oil with donor-distributed soy oil, due to the Israeli destruction of Palestinian olive orchards. The book's author, Laila El-Hadad, found on interviewing Palestinian women, that one of the saddest things regarding the substitutions of ingredients was lack of access to wild plants and herbs, such as *zataar*, as a consequence of Palestinians being fined by Israel for foraging. The traditional staple grains of *frikah*, bulgur and barley have been replaced by less nutritious white flour and rice. Nevertheless, Palestinian women still manage to feed their families with amazing ingenuity, creating traditional recipes from the vegetables and pulses that are available. The determination to continue to cook their traditional recipes is fundamental to displaced populations such as the Palestinians, in an effort to retain their cultural identity. Moreover, the "culture shared in the kitchen is not only food related. The hours spent chopping and stirring often double as a vital cultural space in which women neighbours and family members debate politics and exchange gossip, recount stories and recite poems" (121).

Because Palestinians have been deprived of their land and other aspects of their cultural identity, the issue of culinary cultural appropriation is of particular concern to them. El-Hadad relates how Jewish Israeli chefs have 'discovered' traditional Palestinian recipes and food such as *frikah*, without acknowledging their ethnic origins, and with newspapers and magazines claiming recipes and ingredients such as *hummus, shakshuka* and *zaatar* as 'authentic Israeli food'. As El-Hadad says: "Palestinians take great offense to such attempts at cultural appropriation, viewing them as yet another form of colonization" (47).

The determination of displaced people, including those displaced within their own countries, is illustrated by Romanian cookbook writer Irina Georgescu (2020). In her cookbook '*Carpathia: Food from the Heart of Romania*' she describes how during the twentieth century the communist regime attempted to obliterate Romanian culture by homogenising traditions, including artisanal food traditions such as charcuterie, cheese-making and wine-making. They did this by demolishing villages and forcing the rural population into apartment living. She relates how only one standardised cookery book was allowed. However, the displaced peasantry got around this by creating a black market for traditional recipes which were "passed down by word of mouth or written down on napkins or on loose pieces of paper" (217).

'Relying on Ready-to-Eat Meals': The Loss of Culinary Skills

Women have a complex and often contradictory and ambivalent relationship to cooking, with feelings about it ranging from a passionate interest in and love for it, to hatred and resentment towards it, as well as feelings of fear and inadequacy. Indeed, women often have a mixture of such feelings towards cooking: Shashi Deshpande (1992), writing of her memories of her female relatives in India, describes their complex and often contradictory relationship to cooking and kitchens, being both a source of hard labour and servitude to others, especially men, but also a source of intense pride and female community (although in some households women tyrannise other, usually younger women, often the daughter-in-law). As Dena Attar explains, women's complex and often intense relationship to cooking is explained by the fact that since, historically, "women in general were expected to be able, and willing to cook, by virtue of our sex…it is impossible to feel neutral about cooking" (cited in O'Sullivan 1987, 8). For women cooking is not a value-free activity (as it tends to be for men), but carries with it much 'emotional baggage'; for instance, if women do not like cooking, they tend to feel guilty about it, especially if they have children. As a result, it is "inevitable that food is often problematic for women with the endless work of buying and preparing, and trying to please and provide for others" (83).

Even though few, if any, women in modern Western society do all of the domestic tasks that previous generations of women did—sewing, spinning, weaving, preparation of herbal remedies, etc.—most women still prepare food and cook for others on a regular daily basis. Some women resent this; others enjoy it and find cooking creative and rewarding—even though these women, too, often dislike the constant demands and expectations of having to prepare food day after day, year after year, especially if they also work outside the home, as most women now do (Bunting 2004, 117–118). Women are torn by the conflicting demands on their time and energy and often feel guilty about not having the time to prepare 'proper family meals'. The increased pressure on women doing two jobs—that is, domestic work and paid work outside the home, the latter being often by choice due to its conferring some financial independence, but also because most families now depend upon two incomes—means that there is less time to cook. As a result, there is increasing dependence on processed foods, with the food industry quick to exploit the lack of time women have for cooking. The increasing dependence on processed foods at the expense of wholesome meals cooked 'from scratch' has contributed not only to a decline in women's cooking skills and knowledge, but is also having serious negative health consequences, most notably a rapid increase in obesity and related dietary diseases.

The decline in food knowledge and culinary skills has long been recognised (by some) as a problem. Speaking of the decline in cooking skills in the industrialised world as a result of a decrease in the passing on of food and cooking knowledge from one generation of women to the next, British food policy academics Tim Lang and Michael Heasman (2004) point out that, especially in industrialised societies, the traditional skills base for cooking is changing, and that "the matrilineal thread to food culture may be fraying: as family structures alter, mothers and grandmothers are not the important purveyors of food culture – teaching what to cook and how – they once were" (208). According to British food writer Joanna Blythman (2006a), in societies with healthy food cultures "food appreciation and cooking is an effortless chain of knowledge, passed on in a domestic setting through the generations informally and without fuss,...[with] children learning about food and cooking at home..[by] watch[ing] adults, usually mothers and grandmothers, turning out a family meal almost every day" (206). She argues that in Britain children are no longer being exposed to domestic cooking and therefore no longer absorb food knowledge and cooking skills as they once did. This is because Britain has increasingly become "a 'no time to cook' culture, [where] parents are unable to fulfil this natural function because they themselves have become steadily de-skilled and reliant on ready-made meals" (207). Because children are no longer exposed to home cooking as they once were, they have increasingly come to see industrially processed food as the desirable norm; this is a result of no longer having the experience of eating home-cooked meals as a benchmark against which to judge the industrial product. As she graphically expresses it: "[G]enerations reared on a diet of chemical sweets, bone-rotting fizzy drinks, sugar-coated cereals, sweets, crisps and poultry slurry nuggets, soon find the natural taste of unprocessed foods unpalatable" (257).

As I discuss in Chapter 6, the severing of a connection to land and food production brought about by urbanisation during the Industrial Revolution resulted in a serious erosion of food culture in Britain, with lack of resources and time to prepare nutritious meals being a problem that especially afflicted the working classes. However, what is sometimes referred to as the 'decline of the family meal' is an issue that now affects much of modern industrialised societies across the socio-economic spectrum, especially in the English-speaking countries such as the United States and Britain. Increasingly in these societies unhealthy fast-food takeaways, additive-laden snacks and pre-prepared packaged supermarket meals have come to replace meals cooked using basic raw ingredients. As Blythman points out "[a]s a nation, we have tacitly agreed to outsource the lion's share of food to a profit-driven industry" (Blythman 2006a, 245).

Blythman is scathing of the state of food knowledge and cooking skills in Britain, which, in contrast to other European countries that still have vibrant food cultures, is the "cultural exception…, second only to the US in its capacity to shock outsiders with its eating habits, a country that has well and truly lost the gastronomic plot, a food recidivist, demonstrating precious little capacity for improvement" (2006b). She attributes the absence of a healthy food culture and what she considers the deplorable lack of cooking skills in Britain to several factors, including an ever-increasing reliance on ready-made supermarket meals, which means that less and less actual cooking is taking place; this dependence on convenience foods "has a corrosive effect because the less you cook, the less you are likely to feel able to cook" (2006a, 83).

Caraher et al. (2000) found from their studies that the British spend less time in the kitchen than continental Europeans and that, with regard to cooking, British society is divided between a minority who spend "large amounts of time in the kitchen …[while the majority] are disengaging from the task and relying on ready-to-eat meals and take-out" (28). Blythman contends that the increasing British reliance on supermarket ready meals has led over the past couple of decades to the "modern myth …that it is possible to abandon cooking entirely but still continue to eat great food every day of the week", adding that such an idea would be considered 'preposterous' by countries with healthy food cultures (Blythman 2006a, 78). This is now further compounded by the development in recent years of UberEats—a true product of the neoliberal underpaid gig economy—in which people do not even have to make the effort to go and pick up takeaways, thereby increasing sedentary lifestyles, with the associated health problems, such as obesity.

Family meals are seen as central to family life and much concern has been expressed about the perceived decline of families eating together—that is, sitting around a table for a meal. The average amount of time that was spent on meal preparation had shrunk from one hour in 1980 to thirteen minutes in 2001 (Caraher et al. 2000, 28). Furthermore, results from a survey carried out in 1997 revealed that two-thirds of British families sat in front of the television, rather than around a table to eat, and that by 2004 this figure had risen to seventy-five per cent, a fact reflected in a sharp decline in the sale of dining tables, which "have effectively become redundant".[1] Despite the fact that most people still thought that shared meals sitting at a table was the ideal, Blythman concludes that in reality: "[B]ritain's chosen mode of eating for the 21st century increasingly consists of consuming a reheated factory-made meal on the lap, in front of a flickering screen" (Blythman 2006a, 87). Lack of time is usually considered to be one of, if not the main factor in explaining the decline in home-cooked meals and the increasing reliance on fast food and processed meals. While it is true that a majority of women are now engaged in paid work outside the home, meaning that there are ever-greater demands on their time and energy, some writers point out that this fact has been exploited by the food industry in order to increase sales of industrially processed foods. Laura Shapiro (2005), for instance, contends that the obsession with saving time was a ploy used by the food industry in the years following the Second World War in the United States, whereby the industry created "a manufactured sense of panic which began to pervade everyday cooking", with advertisements "reminding readers again and again how busy they were, how frantic their days, how desperately they needed products and recipes for quick meals" (78). She explains that in the post-war years food manufacturers were determined to apply to civilian life processed food which had been developed for the army. In an attempt to commercialise it on a large scale, the food industry was determined to "persuade millions of Americans to develop a lasting taste for meals that were a lot like field rations" (8).

During the post-Second World War years, the food industry went to great lengths to convince women of the central importance of domesticity to their lives, whilst simultaneously persuading them to use the newly emerging pre-packaged processed foods. The food industry was determined to redefine the very meaning of cooking, to displace the traditional meanings with regard to cooking that had been built up over many generations by creating a 'new' cuisine that was a product of manufacturers' test kitchens. Consequently, "unlike its traditional counterparts, this cuisine was *detached from the usual sources of culinary invention – the weather, the crops, the special occasions for feasting, the daily needs of home cooks*" (55; italics added). According to Laura Shapiro (2005), the food industry had two challenges: one, to educate the American palate to accept pre-packaged foods, and two, to wean women away from their sense of ownership of cooking, or at least to somehow convince them that they were still involved and being creative despite using such products. The first challenge had already been largely met due to the fact that, since the late nineteenth and early twentieth centuries, Americans had become accustomed to the taste of processed foods, "with sweetness the dominating flavor", and as a result had already developed a sense of taste "that tended to perceive imitation as plenty good enough" (57). The second challenge facing the food industry was more difficult and entailed several decades of wearing women down because the "more daunting obstacle that confronted the creators of packaged-food cuisine was that women seemed unaccountably attached to cooking" (60).

Shapiro contends that the resistance to processed foods was based not so much upon their taste, but rather on women's continuing emotional attachment to cooking. Women were initially suspicious of and resented the attempts by the food industry "to elbow [them] away from the stove, and few … welcomed the intrusion" (25). Women's emotional attachment to cooking, one that is intimately bound up with their self-image as homemakers and carers of their families meant that: "For their own stubborn reasons, many women still shared, perhaps even treasured, a powerful understanding of what it meant to cook for loved ones at home, and they had relegated convenience foods to a minor role in meal planning" (55). Although women in the post-war years did use some processed foods, they tended to avoid the ready-made canned and frozen foods being promoted by the food industry. One of the strategies employed by the food industry in their battle to dominate American culinary culture was to assert that women disliked and resented cooking, that women found cooking thankless drudgery and therefore welcomed and were enthusiastic about convenience foods. This was despite evidence to the contrary; indeed, a number of studies at the time found that "women liked cooking better than any other household task, or at the very least disliked cooking less than they disliked other chores" (44).

In addition to the fact that a majority of women actually enjoyed and took pride in preparing meals, being very selective in their use of certain packaged foods that they considered acceptable, and which did not compromise the integrity of their cooking, they also felt guilty if they resorted to pre-packaged meals and ready-made frozen dinners, such as the in/famous TV dinners introduced by Swanson in 1954. Women considered these only acceptable in an emergency, as such a meal did not represent the care and love associated with proper home cooking.

> No matter how appealing the prospect of doing something as utterly labor-free as opening a can of macaroni and cheese, spooning the contents into a pan, heating it up and serving it, very few women were able to convince themselves that they had produced a meal. The moral obligation to cook was simply not satisfied. Where was the personal dimension, the hospitality, the sense of gift from heart and hands that had characterized the notion of home cooking for millennia? (Shapiro 2005, 52)

The American food industry employed every means at its disposal to convert women to pre-packaged foods, including advertisements, strong influence over the food sections of newspapers and magazines, and the creation of fictional cookery experts, such as Betty Crocker. Psychologists were employed by food companies to analyse and devise strategies to overcome women's resistance to convenience foods, to find ways around women's guilt associated with instant pre-packaged meals, including convincing women that they could be creative with them by 'doctoring' them up. The ploy of 'doctoring', which was aimed at women's need to be creative and personally involved in cooking, involved putting one's 'personal touch' to a recipe based almost entirely on prepacked, frozen and canned food. One such example was the 'twenty-minute roast', consisting of wedges of Spam glazed with orange marmalade (Shapiro 2005, 63). Another strategy called 'glamorising', because it suggested 'luxury and sophistication' was the addition of alcohol, such as a splash of wine or sherry to canned soup or flaming an instant dessert with brandy (65). Significantly, the food industry's obsession with saving time did not apply when it came to the 'glamorising'—that is, decorating—of pre-packaged cakes, which could take hours (78). In fact, as Erika Endrijonas (2001) explains, the 'doctoring' and 'glamorising' often involved women spending "enormous amounts of time on silly dishes"; one such example recalled by a 1950s homemaker involved shaping mashed potatoes "into Bartlett pears, with cloves for stems, and glazed with egg whites" (164). Apart from the fact that this makes a mockery of women's desire for culinary creativity, it also gives lie to the industry's contention that they were concerned with saving women time, when in fact their real purpose was to sell more processed and pre-packaged food products.

The issue of time—or lack of it—is undoubtedly very important for women, even more so now that a majority have to juggle two jobs, both inside and outside the home, and also given, in general, men's continuing lack of involvement in domestic tasks. Despite the unquestionable demands on women's time and energy, Blythman (2006a) disputes the popular thinking that people don't have time to cook. She argues that with regard to what she calls the 'no-time-to-cook' culture that prevails in Britain, the problem is largely one of distorted priorities arising from "the deeply-rooted British idea that food is of trivial importance" (71) in the overall scheme of things, and that almost anything else is more worth spending time on. This attitude is reflected in and promoted by the food industry, with one manufacturer of convenience foods commenting that: "People generally are trying to fit more pleasurable things (than cooking) into their lives" (69). Indeed, the food industry likes to promote itself as being a boon to women, with one ready meals entrepreneur in Britain claiming that he had "broken the housewife's shackles from the kitchen" (Blythman 2015, 23).

It is undoubtedly true that one aspect of the debased food culture in Britain (which is a result of the processes that I discuss in the following chapter) is a lack of prioritising of food and cooking as central to the quality of life. However, the 'no-time-to-cook' culture is not simply a problem or failing of the individual—an approach that tends to blame women—but rather is fundamentally a structural or institutional problem. The reality is that in the modern industrialised world people in general, and women in particular, have less and less time to do all that is demanded of them, including and perhaps especially in maintaining the caring work that has traditionally been women's role. This is a situation that prevails particularly in the English-speaking industrialised world, especially in the United States and the United Kingdom, and also increasingly in Australia (Bunting 2004, 302). As Madelaine Bunting explains in her analysis of the impact of overwork in Britain, the technological revolution that occurred during the 1970s promised to reduce working hours and increase people's time for leisure, has in fact resulted in the opposite, with the advent of the internet, emails and mobile phones blurring the traditional boundaries between the workplace and home. As Bunting attests, the American long-hours work culture became established in Britain during the market-driven economic rationalist Thatcher era of the 1980s, with the result that British workers now work the longest hours of any other country in the European Union. The "dramatic acceleration" in British men's work hours is even "outdone by what is happening to women", with the number of women working more than forty-eight hours a week having increased by 52% since 1992, and with one in eight of the female workforce working over sixty hours a week (9).

The overwork culture that now prevails in Britain, and that affects both men and women, but especially women, has resulted in a "care deficit towards those dependent on us – children, the elderly and each other" (210). The overwork culture "demands that we belittle and marginalise the ethic of care", resulting in "depriving [people] of the time and energy to sustain and nurture lives" (211), that is, the caring that is the basis of sustaining human relationships which always has been and still is overwhelmingly women's responsibility. Inevitably, the lack of time and energy caused by the overwork culture in Britain with "women struggling to keep up, stretched, as if on a rack, between the office and the home" results in lack of time for preparation of meals and an increasing reliance on processed foods and supermarket ready-made meals, a situation that is seized upon and exploited by the food industry. The handing over of control to the food industry for people's day-to-day nourishment is not only risky from the point of view of health, but also an example of "the increasing commercialisation and commodification of the domestic sphere" (210–211). As Bunting expresses it, speaking of the increasing commodification of people's personal and emotional lives, the "private world is [being] commercialised, contracted out", representing a "triumph of market capitalism" (306).

Blythman (2006a) asserts that the food industry has a much stronger hold on Britain than the rest of Europe because, as a 2002 report into market 'penetration' found, Britain was "the country where traditional eating habits were the most eroded" (76). As a result, the food industry regards Britain as a particularly lucrative market, further undermining its fragile food culture. The concept of 'time poor-cash rich' consumers is perpetrated and exploited by the food industry, which is anxious to profit from the extra income generated by two-income families. Blythman points out that, ironically, despite the fact that British women are doing less and less cooking, they are still overwhelmingly responsible for providing daily meals, thinking about, planning and shopping for the individual requirements of different family members. These demands have become even more complex with the advent and development of a vast array of microwaveable ready meals, rather than the traditional standard meat and three veg family meal that everyone once ate. Therefore, "deciding what everyone is going to eat each night may be more stressful than simply cooking from scratch. This is a new pressure on women that their female forebears who followed a cyclical repertoire – roast on Sunday, cottage pie on Monday, fish on Friday and so on – never had" (82–83).

There is a long male tradition of blaming women for a real or perceived decline in cooking skills. For instance, a certain Philip Wylie, in a polemic against the increasing use of processed foods, '*Generation of Vipers*' written in 1942, attributed the decline in the standard of home cooking to an 'unholy alliance' between women and big business (Smith 2001, 186). Indeed, the fact that pre-packaged processed foods were directed at the female home cook reinforced the prevailing misogynist view held by some men that women were innately inferior cooks to men.

'A Form of Diversion and Escapism': Cooking as a 'Spectator Sport' and the Decline of Home Cooking

Perhaps more than any other society, in Britain there appears to be an inverse relationship between the amount of time spent on cooking and that spent watching television cooking shows, which have burgeoned as cooking knowledge and skills have declined. As Blythman (2006b) says: "In spite of the plethora of TV cooking programmes, cookery articles in magazines and newspapers, and cookery books,...less and less cooking is being done up and down the land...we seem to spend more time watching chefs cook than cooking ourselves" (53). In their study of the influence of television and celebrity chefs on the cooking habits of the British public, Caraher et al. (2000) found that some people were spending a lot of time in the kitchen, whilst the majority are "disengaging from the task and relying on ready-to-eat meals and take-out", with “cooking…moving from… a production skill to a section of the leisure industry, with a consumer and leisure focus” (28–29). Cooking in the United Kingdom is not a life and health-enhancing life-skill as it is in countries with a still intact food culture, but rather an optional hobby or statement of lifestyle. Indeed, the stereotype is that of a public addicted to television cooking programmes, with viewers eating processed supermarket ready meals, whilst watching a celebrity chef whip up elaborate and exotic recipes, a situation in which cooking is not something people simply do themselves, but which has become entertainment, a voyeur activity, often referred to as ‘gastroporn’.

Nevertheless, Caharer et al. found that, although generally the public view television cooking shows primarily as entertainment, there was nevertheless “a thirst for learning about food and cooking”. They conclude that the two major sources of learning are the family and school cooking classes, which provide the foundation of basic cooking skills upon which to build further skills and knowledge, for instance through reading cookbooks. Without at least basic cooking skills this is, however, extremely difficult, and it is precisely these skills that a large number of people in Britain no longer have. They point out that this explains the huge popularity of Delia Smith’s basic cooking course. However, it was precisely the basic cooking skills being taught by Delia Smith that were usurped by the advent of the usually male celebrity television chef during the 1990s. Whereas Delia Smith’s television cooking shows were about enabling viewers to gain the skills and confidence to cook the recipes she demonstrated, the recipes prepared by the celebrity chefs tend to be elaborate, often based on complicated professional restaurant cooking rather than the home cooking that characterised Delia’s programmes. The cooking of usually male celebrity chefs is generally not so much about making cooking skills accessible to viewers, but tends rather to be an egotistical display of their culinary prowess, often using extravagant and exotic ingredients. Consequently, rather than inspiring confidence in viewers, celebrity chefs often have the opposite effect, with the media-driven cult of the ‘celebrity chef’ probably doing more harm than good with regard to people’s cooking skills, and with the average viewer feeling intimidated and inadequate vis-à-vis the elaborate recipes being demonstrated. The effect of television celebrity chefs, therefore, “is not empowering but paralysing [because they] live in a world glossy with food fashion and busy with perpetual novelty…we let ourselves be entertained by them as a form of diversion and escapism, but we know that all this has little or nothing to do with real life" (Blythman 2006a, 14).

The advent of the male celebrity TV chef with the "efficacy of [the] recipes …considered by broadcasters secondary to his entertainment value" was driven by the perceived need to "appeal to male viewers, to the kind of man who likes to show off his cooking skills at high-profile events but doesn't like to muck in with everyday domestic cooking" (8). According to Blythman (2006a) this 'value-added' approach to cooking shows is yet further evidence of "Britain's traditional lack of conviction that food in its own right merits intelligent interest" (9). Again in contrast to Britain, where food "as a spectator sport has …become a huge industry", in countries such as France and Italy, where food is considered so central to the enjoyment of life, it features far less in the media, with any cooking shows on television having none of the lifestyle hype of the British shows, but "simply demonstrate a recipe and leave it at that" (12).

Although celebrity chefs frequently claim to deplore the poor state of home cooking in Britain, they themselves are implicated in its undermining by their endorsement of and financial connections with the food industry and supermarkets which are so instrumental in causing the erosion of food and cooking knowledge. This is not a new phenomenon: Shapiro describes how, in the years after the Second World War, American food writers, such as James Beard, despite their concern with good food and cooking, were nevertheless, complicit in the promotion of industrialised packaged food through their well-paid endorsement of such products. As a result, "such gestures of support for packaged-food cuisine, coming as they did from some of the best professional cooks in the country, helped erode the foundations of traditional home cooking" (Shapiro 2005, 82). Many of the television celebrity chefs in Britain have strong connections to the food industry, thereby supporting the very institution that is playing such a major role in undermining the foundations of home cooking; examples include Gary Rhodes' financial connections with Tate and Lyle sugar and Jamie Oliver's with Sainsbury's supermarket who, as the "celebrity face of Sainsbury's"(Blythman 2004, 40) gets paid one million pounds a year to feature in the supermarket chain's advertisements (Brindle and Maley 2006), and with even Delia Smith, despite her generally sound approach to teaching home cooking, being handsomely rewarded for endorsing Sainsbury's products. One public health campaign to promote healthier eating and to encourage families to eat together involved a collaboration between Sainsbury's and

Jamie Oliver (who does not shop in supermarkets himself, preferring London's small speciality food shops). This campaign shows just how in thrall to supermarket culture and to the cult of the celebrity chef Britain is. Blythman points out how cooking classes for children that were launched by Sainsbury's in 2003 and promoted as encouraging healthier eating amongst children were, however, also used as an opportunity to promote the supermarkets products with "reminders that if they didn't feel like cooking, they could always get Mum to pick up something ready-made from Sainsbury's" (Blythman 2004, 53).

The steady decline in the learning of cooking in the home environment, that is, the demise of the mostly female intergenerational food and cooking knowledge that was traditionally passed on from mother to daughter, and the failure of schools to teach basic cooking skills due to the shift to food technology geared towards the food industry means that the two most important ways that people have learnt to cook are ceasing to exist. As Blythman despairingly asks, "if [people] don't learn to cook either at home or at school, how are future generations going to be equipped to feed themselves in a way that is independent of the food industry?" (2006a, 216). And as Australian restaurateur and promoter of cooking classes in primary schools, Stephanie Alexander expresses it, the increasing reliance on convenience foods, the decline of family meals, and the ignorance of how food is grown and produced is resulting in "children and young adults who have no connection with the growing cycle of food on the planet ...[and who] without an example of savoury, delicious dishes emanating from the kitchen...grow up only able to feed themselves from packets, jars or cans" ("Who's cooking now?" *The Age*. Melbourne, October 9, 2001).

'The Epitome of Patriarchal Oppression': Feminism and Cooking

In view of women's often subservient role within the family, one which demands self-sacrifice, self-denial and constantly putting the needs of others first, it is hardly surprising that feminists should emphasise the need for women to quit the private undervalued unpaid domestic sphere for the more valued male-dominated public domain. Simone de Beauvoir (1953), for instance, lamenting the lot of the 'wife-servant', stated that "Few tasks are more like the torture of Sisyphus than housework, with its endless repetition...It is tiresome, empty, monotonous, as a career. Eating, sleeping, cleaning – the years ...spread out ahead, grey and identical. The battle against dust and dirt is never won" (438–440). De Beauvoir's contention that "woman's work within the home...is not directly useful to society, it does not open out on the future, it *produces nothing*" (440, italics added) reveals, despite her many important feminist insights into women's situation within patriarchal society, a masculinist mindset that places little or no value on women's unpaid labour within the domestic sphere. As feminist economist Marilyn Waring(1988) points out in her important book *Counting for Nothing: What Men Value and What Women are Worth* on the lack of recognition of women's unpaid work in international accounting systems, the view that women, housewives and mothers are 'non-productive' is both a result of and contributes to the invisibility of women's massive unpaid contribution to the world economy.

Despite her dismissal of women's household productivity, de Beauvoir did, however, concede that the "preparation of meals, is work more positive in nature and often more agreeable than cleaning" because of its potential for creative expression and because it entailed expeditions to the market, "often the bright spot of the day" (439). Indeed, de Beauvoir almost waxes lyrical when she talks of cooking (explained in part, no doubt, by her being French and therefore a product of a society with a strong food culture and a lively interest in food). Asserting that the invention of gas and electricity have destroyed the magic of cooking with fire, she says:

> With her fire going, woman becomes a sorceress; by a simple movement, as in beating eggs, or through the magic of fire, she effects the transmutation of substances: matter becomes food. There is enchantment in these alchemies, there is poetry in making preserves;...Cooking is revelation and creation; and a woman can find special satisfaction in a successful cake or a flaky pastry, for not everyone can do it: one must have the gift. (de Beauvoir 1953, 439)

This sentiment is echoed by English food writer, Jane Grigson, who succinctly expresses it: "Intelligent housewives feel they've a duty to be bored by domesticity. A fair reaction to dusting and bedmaking perhaps, but not, I think, to cooking" (2007, 9). It has become fashionable to blame feminism for the crisis that exists in what is called the work-life balance: women, who are exhausted and overstretched by the competing demands of their paid work and work in the home, are frequently accused of creating this situation by "wanting to have it all", that is, a paid job or career as well as children and a family life, a dilemma that does not generally apply to men. As Madelaine Bunting (2004) points out, "the phrase 'work-life balance' tends to individualise what is essentially a structural problem of the way society is organised, as well as placing the onus on women in particular, who often feel personally inadequate for failing to achieve a balance in their work and home lives." (214) She argues that the "feminisation of the issue" perpetuates the misogynist tradition of attributing social problems to a "problem of female character", when in fact:

> What is at stake here is not women's strength of character, but the market economy on human sustainability: the reproduction of daily life, the routines of daily life that meet our physical and emotional needs, which are by far the biggest part of personal well-being, and which gives rise to and reinforces the relationships of interdependence within which people flourish. (215)

The tendency to blame feminism for the crisis in 'work-life balance' not only provides an opportunity to continue the misogynist tradition of blaming women, but also represents a fundamental misunderstanding of the feminist radicalism of the 1970s which recognised that "the entire way we worked had to be revolutionised" (2004, xviii), insisting that it was not only women's lives that need to be transformed, but men's too. Whilst women have made inroads into the traditional male workplace, albeit still generally on unequal terms, the sexual division of labour within the private sphere of the home has remained largely unchanged; women still overwhelmingly bear the burden of domestic work, even when they work outside the home, with the notable exception of the more desirable aspects, namely some men's occasional stints of cooking, as discussed in the previous chapter. In their analysis of heterosexual power relations within the family, sociologists Christine Delphy and Diana Leonard (1992) argue that married women "pay for the 'freedom' to work outside the home and to have 'an independent income' (that is, one over which they have some limited rights and which could support them should the marriage break down) with a double day's work" (118). They ask the interesting and important question of whether women's entry to the work market "has now reached a threshold where there is a real conflict between patriarchy (that is, the patriarchal family) and capitalism for women's labour, answering that, in their opinion, it has, with the result being that women are "working longer hours: working for both capital and their husbands" (121). Bunting suggests that the main barrier to men's lack of involvement in housework is their long working hours. This is no doubt a factor, and one which requires a fundamental change of priorities in the 'overwork culture', which Bunting says needs to be insisted upon by men even more than by women (2004, 264). Nevertheless, studies such as those cited in Chapter 4 indicate a continuing resistance on the part of men in general to shoulder more of the domestic burden.

Many younger women, when they realise that it is extremely difficult, if not impossible, to successfully combine having children and managing the household with having a fulfilling career, feel disillusioned and even betrayed by what they believe was the promise of feminism. But as Natasha Campo (2005) points out, the myth of being able to "have it all" was not one that was generated by second-wave feminism, but by the popular media. Moreover, working-class women, who have always had to combine child-rearing and domestic duties with paid work, have always seen themselves as "doing it all", rather than "having it all" (72). Of course, it is easier and less threatening for women to blame their feminist mothers for their not being able "to have it all", rather than direct their frustration and anger towards the real problem—namely, the patriarchal work structures that are still "built around the needs and lives of men with stay-at-home wives" (65) and the continuing resistance of most men to taking on a fair share of household work.

Some female commentators outspokenly defend, and even applaud, the lack of cooking skills, especially amongst younger women, as a sign of feminist progress, equating the 'daily grind' of cooking with cleaning the toilet, and with one rather absurd and flippant comment being: "Nobody these days wants to be a peasant, and girls have better things to do than pluck chickens" (cited in Blythman 2006, 80–81). One Irish writer, speaking of her experience as an 'au pair' in France in the 1960s, recalls her horror at the amount of time women spent shopping for and preparing meals, with the "delights of French cuisine [being] produced at the cost of some poor female skivvying away for hours over a hot stove" (Kenny 2004).

Kenny states that she is not opposed to good cooking, but rather that she objects to the burden that she considers it imposes on women;

in her opinion, good cooking is something that should be enjoyed in restaurants, or "as a hobby at home in which either men or women may voluntarily indulge". She applauds the existence of supermarket ready meals as liberating women from the 'tyranny' of the kitchen: "...bring on the fresh frozen lasagne...: heat up the pre-packaged vichysoisse, and take that crème caramel out of its divine Marks and Spencer wrapping" (Kenny 2004). The British are huge consumers of ready meals, with demand growing by 70% between 1994 and 2004, and with 900 million pounds being spent on them in 2004, a rate that is growing at 6% per year (Blythman 2006a, 98). As Blythman (2004) explains, whereas such meals are rarely eaten in the rest of Europe, being generally of a higher standard there and only eaten as an emergency back-up, in Britain they are almost daily sustenance for many people and with the more 'sophisticated' versions of it even considered to be a gastronomic treat. As a result, for many Britons the "idea of a gourmet night is eating a supermarket ready meal on a tray while watching a procession of celebrity chefs cook fantasy food on TV" (52). One of the main strategies of the food industry is to propagate the idea that factory-made food is as acceptable as home-made food, and this has been made possible by the fact that consumers in Britain are "poorly educated in food, de-skilled in cooking, and traditionally indifferent or even resistant to the idea that food is important in life" (Blythman 2006a, 103).

Barbara Haber, curator of books at Radcliffe's Schleshinger Library, says that she often finds herself having to defend the existence of the library's collection of cookbooks to feminists who consider the kitchen "to be the epitome of patriarchal oppression" (Haber 1997, 65). She points out, however, that when she has carefully explained that such a collection represents an important part of women's history, her critics tend to be more understanding and accepting. Although second-wave British and American feminists have tended to see cooking as symptomatic of women's domestic exploitation and oppression and therefore to be dispensed with at the first opportunity, some are now questioning the wisdom of this. Rosie Boycott, a co-founder in the 1970s of the influential British feminist magazine, *Spare Rib,* for instance, considered "that cooking was a demeaning pursuit for women who wanted to get on in a man's world...[with] food preparation [being] so entwined with the role of housewife that we consigned it to the bin of history..." (Boycott 2007). However, she now realises that, in retrospect, the consigning of cooking "to the bin of history" was possibly mistaken, given the decline in people's ability to cook, the increasing reliance on takeaways and supermarket ready meals, and the accompanying obesity epidemic. She considers, correctly, that although feminism is not to blame for this state of affairs, nevertheless, the admonition against cooking by feminists can be seen as "a small link in [that] chain" (2007).

Women's relationship with food and cooking is complex, and, for many, intimately bound up with their sense of identity as women. As Kate Cairns and Josee Johnston (2015) found out from their extensive research, issues to do with food are of particular importance to women. While food preparation and cooking is often felt as a burdensome and compulsory chore, for many women it is also a source of "*positive* emotions of pleasure and creative expression, excitement and empowerment, political passion and familial pride" (160; original italics). They contend that recognition of the sense of pleasure and power that women often derive from cooking, far from being a trivial matter, is in fact important to a feminist understanding, as it "involves long-standing struggles to reclaim the domestic sphere as a site of value, and not just a realm of degradation that women should flee once they are 'liberated'" (163). Writing about her mother's enjoyment and pride in her cooking, Gloria Wade-Gayles, says that for Black women like her mother, food and cooking was of central importance to their identity, and a sense of emotional and cultural connection to their foremothers who had been slaves on white plantations: "Cooking was the centrepiece of their bonding…Women…went to the kitchen to work, to serve, to think, to meditate, and to bond with one another…In their kitchens, the women experienced influence, authority, achievement, and healing" (Wade-Gayles 1997, 97). Cairns and Johnston also found that women's valuing of the activity of cooking often extended to issues of caring for others and for the planet. Significantly, a valuing of cooking, and its compatibility with feminism is cited by women of diverse immigrant backgrounds, such as Yin who describes how women's food work is valued in countries such as China. Another interviewee, a food activist, Debbie, states that "…cooking from scratch was part of my liberation and it's part of my connection to generations of my foremothers, so I don't have to boycott home-cooking as a feminist" (Cairns and Johnston 2015, 167). Meredith Abarca (2006) makes an interesting distinction between the kitchen as woman's 'place' and woman's 'space': 'place' carries 'patriarchal ideological baggage' (220) that associates the kitchen as a site of women's exploitation and subordination, whereas 'space' describes the kitchen as a place that women feel free to claim as their own and to express creativity and agency. She gives the example of a working-class Mexican American woman who felt oppressed and depressed when having to share a kitchen with her mother-in-law and sisters-in-law who were unappreciative of her work and critical of her cooking. However, when she eventually got a kitchen of her own, she felt liberated and in control and able to make it her own 'space'.

Although it is certainly true that food preparation and cooking is intimately and intrinsically bound up with women's mandatory role of serving and servicing others, namely men and children, to reject it entirely is a mistake. This because cooking is important from the point of view of health, as well as providing an important connection to the environment, the natural world and seasonal cycles, and also because of its potential for creative satisfaction. Indeed, Cairns and Johnston (2015) make the important point that the demeaning of cooking as something to be rejected in the interests of women's liberation is, in fact, contrary to a feminist ethic. In relinquishing this and handing it over to either men and/or the food industry, women hand over one of the few traditionally female areas that has the potential for creative satisfaction and a sense of control or authority. Ynestra King warns of the dangers to women of rejecting all aspects of being female in their quest for equality with men, with the risk of "throwing out the baby with the bath water", and that in doing so, "in a sense they have given away what little power women as a group have had without receiving an equal share of male power and privilege, however it might be defined" (King 1990). When feminists regard cooking as unimportant or simply a source of women's oppression, they run the risk of rejecting one of the few areas that have been a source of potential power as well as pleasure. Rather than reject what is an important and potentially pleasurable, rewarding and even empowering aspect of life, feminists need to be demanding a complete re-evaluation of priorities—as indeed many, radical feminists in particular, are. Feminist philosopher, Carol McMillan (1982) points out that, in viewing as 'inherently inferior' the private realm and those activities associated with it, such as cooking, feminists run the risk of subscribing to the same evaluation as male supremacist thinking, which considers activities done by women, such as domestic cooking, to be less important than those of men's.

Despite women's complex relationship to food responsibilities, whether viewed as oppression and exploitation or as an empowering activity, "either way, women's work with food ties them to their families and friends, and in some cases maintains cultural traditions that are at the heart of women's identities" (Sachs et al. 2016, 26). With regard to women experiencing a sense of control and power through food, in a gendered analysis of food and power, cultural anthropologist, Carole Counihan makes an important distinction between masculine use of power and that of women: "[A]lthough control of food can be the strongest weapon of coercion, for women it is not. In no culture is it acceptable for women to deny food to their families, whereas it *is* acceptable for politicians – mostly male – to deny food to entire populations for political ends" (Counihan 1988, 53; original italics). An example of the masculine exercise of power with regard to food is the withholding of food during war from the enemy nation in order to starve the population into submission.

In this chapter I have argued that women, and feminists in particular, should understand cooking not only as a source of oppression and exploitation, but as an essential part of human culture which has largely been developed by women. Consequently, women need to take pride in this, and not let men get away with their smug and arrogant claim to have discovered cooking and the nonsense that men are naturally better cooks. As Marilyn Frye (1983) states in her important analysis of patriarchal construction of reality, male supremacy's ability to define what is real means that men can make such unfounded claims and have them accepted as fact. Women need to see through this lie, and to not naively agree that men are innately better cooks, or to be unquestioningly grateful to men when they choose to only do the more enjoyable and creative household tasks, namely cooking.

Cooking can only flourish in a healthy food culture—that is, a society that respects and protects the rural base of food production; supports and encourages small-scale sustainable agriculture, markets selling seasonal produce and small specialist food shops; and, also importantly, one that gets the work-life balance correct, that is, prioritises the quality of life over long work hours. Time is required for sustainable living, whether it be for maintaining relationships, the growing and preparing of food and the leisurely sharing of meals, all of which are essential to community and human well-being. The opposite is now the case in much of the industrialised world; there is less and less time for these life-enhancing activities due to the insatiable demands capitalist production makes on workers, and increasingly on women who still have to unfairly bear the double burden of work both inside and outside the home. As Madeleine Bunting (2004) argues, the work ethic needs to be replaced by an ethic of caring, in which—citing the historian E. P. Thompson—"people might have to relearn some of the arts of living lost in the Industrial Revolution" (22). In the following chapter, I examine the development of the industrialisation of food systems and the detrimental effects of this on women's traditional food knowledge and skills.

Note

1. It is quite probable that this situation has changed again—with many families, especially teenagers, eating their meals while watching online streaming services on their devices.

References

Abarca, Meredith, E. 2006. *Voices in the Kitchen: Views of Food and the World from Working-Class Mexican and Mexican-American Women*. College Station, TX: Texas A&M University.

Amad, Abla. 2001. *The Lebanese Kitchen*. Melbourne: Viking.

Attar, Dena. 1990. *Wasting Girls' Time: The History and Politics of Home Economics*. London: Virago.

Azzam, Itab, and D. Mousawi. 2017. *Syria: Recipes from Home*. London: Orionbooks.

Birdwell-Pheasant, Donna, and Lawrence-Zuniga, Denise, eds. 1999. *House Life: Space, Place and Family in Europe*. Oxford and New York: Berg.

Blythman, Joanna. 2004. *Shopped: The Shocking Power of British Supermarkets*. London: Fourth Estate.

Blythman, Joanna. 2006a. *Bad Food Britain: How a Nation Ruined its Appetite*. London: Fourth Estate.

Blythman, Joanna. 2006b. "Can't Cook, Won't Cook". *Observer Food Monthly*. London.

Blythman, Joanna. 2015. *Swallow This: Serving Up the Food Industry's Darkest Secrets*. London: Fourth Estate.

Boycott, R. 2007. "Why a Woman's Place Is in the Kitchen." *The Guardian*. London.

Brindle, D., and J. Maley 2006. "Jamie Oliver in Talks Over Campaign for Family Meals." *The Guardian*. London.

Brissenden, Rosemary. 2007. *South East Asian Food*. Singapore: Periplus.

Bunting, Madeleine. 2004. *Willing Slaves: How the Overwork Culture Is Ruling Our Lives*. London: HarperCollins.

Cairns, Kate, and Josee Johnston. 2015. *Food and Femininity*. London and New York: Bloomsbury.
Campo, N. 2005. "'Having It All' or 'Had Enough'? Blaming Feminism in the *Age* and the *Sydney Morning Herald*, 1980–2004." *Backburning, an Issue of the Journal of Australian Cultural Studies* 84: 63–72.
Caraher, M., Tim Lang, and Paul Dixon. 2000. "The Influence of TV and Celebrity Chefs on Public Attitudes and Behaviour Among the English Public." *Journal for the Study of Food and Society* 4 (1) (Spring): 27–46.
Carney, Judith A., and Richard N. Rosomoff. 2009. *In the Shadow of in Slavery: Africa's Botanical Legacy in the Atlantic World*. Berkeley and London: University of California Press.
Contini, Mary. 2002. *Dear Francesca: A Family Memoir of Life, Love and Cooking*. Sydney and Auckland: Bantam.
Counihan, Carole. 1988. Female Identity: Food and Power in Contemporary Florence. *Anthropological Quarterly* 61 (2): 51–62.
De Beauvoir, Simone. 1953. *The Second Sex*. London: Jonathan Cape.
De Silva, Cara, ed. 1996. *In Memory's Kitchen: A Legacy from the Women of Terezin*. Translated by Bianca Steiner Brown. Northvale, NJ: Jason Aronson Inc.
Delphy, Christine, and Diana Leonard. 1992. *Familiar Exploitation: A New Analysis of Marriage in Contemporary Western Societies*. Cambridge: Polity Press.
Deshpande, Shashi. 1992. "Of Kitchens and Goddesses." In *Loaves and Wishes: Writers Writing on Food*, edited by Antonia Till. London: Virago.
El-Hadad, Laila, and M. Scmitt. 2012. *The Gaza Kitchen: A Palestinian Culinary Journey*. Charlottesville, VA: Just World Books.
Endrijonas, E. 2001. "Processed Foods from Scratch: Cooking for a Family in the 1950s." In *Kitchen Culture in America: Representations of Food, Gender and Race*, edited by Sherrie Innes. Philadelphia: University of Pennsylvania.
Field, Carol. 1997. *In Nonna's Kitchen: Recipes and Traditions from Italy's Grandmothers*. New York: HarperCollins.
Frye, Marilyn. 1983. *The Politics of Reality: Essays in Feminist Theory*. Trumansburg, New York: The Crossing Press.
Fulton, Margaret. 1998. *A Passionate Cook*. Sydney: Lansdowne.
Georgescu, Irina. 2020. *Carpathia: Food from the Heart of Romania*. London: Frances Lincoln.
Giard, Luce. 1998. "Doing Cooking." In *The Practice of Everyday Life, Vol. 2: Living and Cooking*, edited by M. De Certeau, Luce Giard, and Pierre Mayol, 149–247. Minneapolis: University of Minnesota Press.
Grigson, Jane. 2007. *Good Things*. London: Grub Street.

Haber, Barbara. 1997. "Follow the Food'. In *Through the Kitchen Window: Women Writers Explore the Intimate Meanings of Food and Cooking*, edited by Arlene Voski Avakian, 65–74. Boston: Beacon Press.

Heldke, L. M. 1992. "Foodmaking as a Thoughtful Practice." In *Cooking, Eating and Thinking: Transformative Philosophies of Food*, edited by D. W. Curtin and L. M. Heldke. Bloomington: Indiana University Press.

Kalla, Joudie. 2018. *Baladi: Palestine—A Celebration of Food from Land and Sea*. London: Jacquie Small.

Kapsalis, Terri. 1997. "Yiayia's Hands." In *Taste Nostalgia*, edited by Allen Weiss, 27–32. New York: Lusitania Press.

Kenny, M. 2004. "La hate cuisine: French Food Always Came at the Expense of Women's Indentured Labour." *The Guardian*. London.

King, Y. 1990. "Healing the Wounds: Feminism, Ecology and Nature/Culture Dualism." In *Reweaving the World: The Emergence of Ecofeminism*, edited by I. Diamond and G. F. Orenstein, 18–28. Orenstein. San Francisco: Sierra Club Books.

Lang, Tim, and Michael Heasman. 2004. *Food Wars: The Global Battle for Mouths, Minds and Markets.* London and Sterling, VA: Earthscan.

Mars, G. A. V. M. 2000. "Food History and the Death of Memory." *Oxford Symposium of Food and Cookery*, St. Anthony's College. Oxford: Prospect Press.

McCann, James, C. 2010, *Stirring the Pot: A History of African Cuisine.* London: Hurst & Company.

McGinniss, Lorraine. 2005. *Feasts and Friends: Women and Food from Across the World*. Fremantle, Australia: Fremantle Arts Centre Press.

McMillan, Carol. 1982. *Women, Reason and Nature: Some Philosophical Problems with Feminism.* Oxford: Basil Blackwell.

Oates, Teresa, and Angela Villella. 2011. *Mangia!Mangia!: Authentic Italian Food Rituals and Family Recipes.* Melbourne: Lantern (an imprint of Penguin Books).

O'Sullivan, S., ed. 1987. *Turning the Tables: Recipes and Reflections from Women.* London: Sheba Feminist Publishers.

Roden, Caudia. 1986. *A New Book of Middle Eastern Food.* Middlesex, UK and Melbourne: Penguin; New York, ON: Viking; Auckland: Penguin.

Sachs, Carolyn E., Mary E. Barbercheck, Kathryn J. Brasier, Nancy Ellen Kiernan, and Anna Rachel Terman. 2016. *The Rise of Women Farmers and Sustainable Agriculture.* Iowa City: University of Iowa Press.

Shapiro, Laura. 2005. *Something from the Oven: Reinventing Dinner in 1950s America.* New York: Penguin Books.

Smith, C. H. 2001. "Freeze Frames: Frozen Foods and Memories of the Postwar American Family." In *Kitchen Culture in America: Popular Representations of Food, Gender and Race*, edited by Sherrie Innes, 175–210. Philadelphia: University of Pennsylvania Press.

Supski, Sian. 2005. "'We Still Mourn That Book': Cookbooks, Recipes and Foodmaking Knowledge in 1950s Australia." *Backburning, an issue of the Journal of Australian Cultural Studies* 84: 85–94.

Sutton, David E. 2001. *Remembrance of Repasts: An Anthropology of Food and Memory.* Oxford and New York: Berg.

Sutton, David E. 2014. *Secrets from the Greek Kitchen: Cooking, Skills, and Everyday Life on an Aegean Island.* Oakland: University of California Press.

Swinbank, Vicki A. 1995. "A Brief History of the Development of Italian Cuisine in Australia." *Journal of the Italian Historical Society* 13 (2) (July–August): 13–16.

Theophano, Janet. 2002. *Eat My Words: Reading Women's Lives Through the Cookbooks They Wrote.* New York: Palgrave.

Wade-Gayles, G. 1997. "'Laying on Hands' Through Cooking: Black Women's Majesty and Mystery in Their Own Kitchens." In *Through the Kitchen Window*, edited by A. V. Avakian, 95–103. Boston: Beacon Press.

Waring, Marilyn. 1988. *Counting for Nothing: What Men Value and What Women Are Worth.* Wellington: Allen and Unwin.

Williams, B. 1984. "Migrant Women Feed Husbands Tamales: Foodways as a Basis for a Revisionist View of Tejano Family Life." In *Ethnic and Regional Foodways in the United States; The Performance of Group Identity*, edited by Linda Brown and Kay Mussell 113–126. Knoxville: The University of Tennesse Press.

CHAPTER 6

A History of the Industrialisation of Food and Its Impact on Women

Industrialisation of the food system over the past century or two, and especially since the Second World War, has had an extremely destructive effect on both the health of people and the environment, as well as contributing to the erosion of food knowledge and skills, an area of traditional female expertise. This process began in Britain with the displacement and impoverishment of much of the rural population that occurred with the enclosure of the commons. That coupled with the Industrial Revolution resulted in the virtual extinguishing of a food culture, with a severing of a connection between the land and the mass of the population, which in turn led to the loss of 'cultural memory', particularly with relation to food. This is now affecting much of the modern world, threatening the cultural food memory of many societies and the biodiversity that sustains them. Cultural food memory, which is vital to all of humanity, is one that has largely been developed and sustained by women over many centuries and millennia, and which is represented by the world's innumerable regional cuisines and food cultures. (This is a topic that I examine in Chapter 8.) It is instructive to examine the process of the erosion of food culture in Britain, and specifically England, as a result of industrialisation as this is a process that is being repeated in various forms throughout the world, and an understanding of this model/process helps us understand similar developments elsewhere.

V. A. Swinbank, *Women's Food Matters*,
https://doi.org/10.1007/978-3-030-70396-7_6

The popular perception of British cooking, and specifically English cooking, is that it has always been bad, that cooking badly is somehow intrinsically part of being British or English. However, English cooking has not always been inferior; indeed, according to food historian, Colin Spencer, during the "first quarter of the nineteenth century [English cooking] won international gastronomic approval" (Spencer 2002, 244). He points out that "every cuisine has its roots within the peasant traditions", and the uprooting of the English peasantry caused by the enclosures resulted in "severe rural poverty and deprivation, essentially truncating the heritage of rural cooking, a loss that at the time no one noticed or thought worth commenting on" (2002, 208). In examining the reasons for the decline in British cooking over the past two hundred years or so, Spencer cites the land enclosures as one of the main influences on the undermining of what had once been a rich British culinary culture. The enclosure of the commons, which both preceded and accompanied the Industrial Revolution, resulted in the destruction of the British/English peasantry, with the "great decline of the English peasantry [having] already happened by the 1730's" (Berg 1994, 84). The loss of common access to land happened in Britain from the fifteenth century onwards with the development of capitalist agriculture, that is, an agriculture whose primary purpose is the accumulation of wealth. This model now dominates the modern world, with the commons under threat being not only rights to land but also to genetic and cultural diversity, topics that I address in Chapter 7.

The Enclosure of the Commons and the Decline of British Food Culture

What was lost in Britain in general, and in England in particular, as a result of the destruction of a peasant-based cuisine was a food culture—that is, a particular way of life which has its roots in the land itself and which informs a whole way of thinking about food: seeing food and cooking not as peripheral, but as central, to one's existence and to the enjoyment and enhancement of life. Furthermore, this enjoyment is not simply self-indulgence, but one which values the importance of all those things that are the basis of and contribute to the flourishing of a food culture, and by extension the care of and respect for the land and the environment and people who produce the food in a sustainable way. The view that food is not of central importance to life, but merely fuel to keep the body going is exemplified by an incident described by anthropologist, David Sutton, who recalls how an Oxford don, incredulous at Sutton's academic interest in food and memory, asks "Food and memory? Why would anyone want to remember anything they had eaten?" Sutton comments that, significantly, whilst "the Oxford don found my topic absurd, ...Greek and Italian colleagues instantly related to it..." (Sutton 2001, 1, 3). While obviously not all English or British people have such a dismissive approach to food as this particular Oxford don, such incredulity is, I would suggest, very much a product of a culture that places relatively little value on food and cooking. Certainly, as Sutton intimates, this is an attitude that is far more intrinsic to English/British culture, than to say, French or Italian culture.

Agrarian historian, G. E. Mingay, notes that "[T]owards the end of the nineteenth century...there was a growing public concern with the decline of the English peasantry" with the number of small owner-occupiers, having "fallen to a very low ebb" (Mingay 1968, 78). In remarking on the contrast with continental Europe where there was still a strong tradition of small owner-occupiers, he suggests that "it would certainly be instructive to examine the reasons why small farmers have survived in much greater numbers in continental Europe, and to consider what relevance these may have for the English experience". Mingay does not say what relevance this had for the 'English experience', simply saying that addressing it "will not be a light undertaking" (32). It is, nevertheless, an important question, and of direct relevance to the question of the decline of English/British cooking, or perhaps more accurately—a vibrant and sustainable food culture.

The decline of such a culture in Britain can be directly linked to the transition from subsistence agriculture to one based on the accumulation of profit, that is, a capitalist one. This process began as early as the late fifteenth century in England, gathering pace during the sixteenth century with the enforced closure of the common fields and grazing lands to which peasant farmers had long had the right of use. As the price of wool rose during Tudor times large estate owners seized common land, mainly in the south of England, for sheep farming or 'sheep walks', as they were called. This was the beginning of the dispossession of the English peasantry, with the "major decline of small owners and of small farmers in general...occur[ing] ...probably between 1660 and 1750" (1968, 31). During the reign of George III, when the increased price of wheat and corn induced farmers to increase production, "more than three million acres was filched from the people" (Drummond and Wilbraham 1958, 174). The agricultural revolution of the eighteenth century that both preceded and occurred concurrently with the Industrial Revolution introduced the beginning of the modernisation of agriculture which viewed small-scale peasant farming as backward and inefficient. One of the main ways in which the small farmer was dispossessed to make way for large-scale capitalist farming was by raising rents beyond the means of small farmers. As a result, "the Georgian enclosures brought desolation to the villages. Everything was lost, the arable land, the grazing for the cottagers' cows, the scrub where they collected their winter fuel, sometimes even their little garden plots…Small farms were being swallowed up by the big estates and many a cottage with its little garden was ruthlessly 'liquidated'" (1958, 176, 280).

Indeed, entire villages disappeared under the impetus of forced enclosure, so that by the middle of the nineteenth-century England "presented a unique and amazing spectacle to an enquiring foreigner: it had no peasants" (Handy and Fehr 2011, 46–47). Historian, Maxine Berg points out that the main changes in agrarian structure were the decline of the peasantry and the rise of a large-scale capitalist farming on a landlord–tenant basis. While the size of the agricultural population shrank, English agricultural output grew by 60–100% between 1600 and 1800 (Berg 1994, 78). This growth in output was due to developments in a modern scientific approach to agriculture and to an increase in the size of farms as a result of land enclosure, with "the winners among the contenders for England's agricultural surpluses in the eighteenth century, (being) neither farmers nor labourers, but landlords" (85). Indeed, the increase in wealth derived from the development of capitalist large-scale farming simply increased the pressure to enforce closure of the commons; between 1700 and 1790 land values almost doubled, and over 4,000 Parliamentary Acts of Enclosure were passed between 1750 and 1850 (Spencer 2002, 208). As a result, "enclosure turned land into absolute private property…[with the] ruling classes, the nobility and the new rich [having] now absorbed most of the fertile land of England…" (208). While the landlords of large farming estates became wealthy, building stately homes, having servants and indulging in conspicuous consumption, the previously often self-sufficient smallholder became dispossessed, and increasingly dependant on intermittent wage labour. Pointing out the close connection between the growth of a wage-based market economy and the increase in poverty, a contemporary commentator observed that:

> The domestic economy of the whole village was radically altered. No longer could the peasant derive the necessaries of life from the materials, the soil, and the resources of his own countryside. Money, which in the sixteenth century had played merely a marginal, though necessary part, now became the one thing necessary for the maintenance of life. Every hour of work now had a money-value, unemployment now became a disaster, for there was no piece of land the wage earner could turn to. (cited in Berg 1994, 85)

Enclosure of the commons, which resulted in the dispossession of the means of livelihood for the English peasantry, laid the groundwork for the emergence of the Industrial Revolution in the late eighteenth and nineteenth centuries. In her fascinating and important book '*The Origin of Capitalism: a Longer View*', economic historian Ellen Meiksins Wood argues that far from the common view that capitalism is intrinsic to human nature, just waiting for the right conditions to flourish, the origin of capitalism, in fact, lies in a specific situation that was unique to early modern England. The background to the development of English industrial capitalism was the development of agrarian capitalism. As a consequence of the enclosures, in England most agricultural land was owned by wealthy landlords, with the land being worked by tenant farmers. This situation contrasted with, for instance, France where the land was owned by small-scale peasant proprietors who were exploited by the aristocracy via taxation. Whereas in France it was not in the interests of the aristocracy to dispossess the peasantry, as the taxation on them was the source of aristocratic wealth, in England during the sixteenth and seventeenth centuries landlords came to view the increase in productivity from their estates as crucial to increasing their wealth. As a result, a market-mediated relationship developed between English landlords and peasants, with the peasantry being increasingly pressured to produce more for the market, whilst being subjected to ever-increasing rents. The result was that the more "competitive farmers would have increasing access to even more land, while others lost access altogether" (Meikens Wood 2017, 101).

With the development of capitalist agriculture the impetus for increased productivity and hence profit became the driving force. Peasant society had "since time immemorial employed means of regulating land use in the interests of the village community" (2017, 107), including providing for the community's less fortunate members, and also customary rights to the use of the commons for grazing rights and fuel gathering, plus rights to post-harvest gleaning on private land. However, with the growth in agrarian capitalism starting in the sixteenth century, "there was growing pressure to extinguish customary rights that interfered with capitalist accumulation" (108). Instrumental to this was the redefinition of property rights, resulting in enclosure. It is important to note that "enclosure meant not simply a physical fencing of land but *the extinction of commons and customary use rights on which many people depended for their livelihood*" (108; italics added). Contrary to the common belief that capitalism was the inevitable outcome of market-based urban societies, "capitalism, with all its very specific drives of accumulation and profit-maximization, was born not in the city, but in the countryside, in a specific place, and very late in human history" (95). As Meikens Wood explains, there had been many market-based urban societies throughout history, including in Asia, the Middle East and in European city-states, such Venice and Florence, but none of them developed a capitalist economy. Despite being great trading centres, commercial activity was based on barter and exchange ('buying cheap and selling dear'), in contrast to the capitalist imperatives of competition and accumulation. Capitalism was based not on "a simple extension or expansion of barter and exchange *but a complete transformation in the most basic human relations and practices, a rupture in age-old patterns of human interaction with nature*" (95; italics added). Indeed, the capitalist approach to agriculture as described by the *Economist* (founded in 1843), "required that land be treated like any other commodity, (and) for this to occur, the primary requisite was the complete victory of enclosure" (cited in Handy and Fehr 2011, 55).

The dispossession of the peasantry's means of subsistence caused by the enclosures created massive social dislocation. The displaced rural population was no longer able to provide for its own subsistence, and was now dependent for the first time in history on waged labour for basic needs. This of course fitted nicely with the development of the Industrial Revolution with its insatiable need for cheap labour, whilst at the same time creating a ready market of the new urban working classes' need to purchase the necessities of life, including food. Indeed, the proponents of enclosure and capitalist agriculture argued that it was necessary to prevent the independence of the rural population in order to create a class dependent on the low-waged labour needed for the new industries. It was considered by such proponents that traditional rights of access to the commons and other customary tenure needed to be extinguished in order to force people into waged labour, with hunger being the chief incentive in achieving this. As Arthur Young, Secretary to the Board of Agriculture in the 1790s, expressed it: "Everyone but an idiot knows that the lower classes must be kept poor or they will never be industrious" (Handy and Fehr 2011, 52).

The Impact of the Enclosures on Rural Women

Both rural men and women were seriously affected by the processes of the enclosures. However, women were particularly adversely affected, especially in the loss of their ability to produce food, which formed the basis of their and their families' livelihoods. The general historical invisibility of women, especially poor, peasant and working-class women, and the relative lack of historical records of small-scale peasant-based food production and preparation make an accurate assessment of these questions difficult. Nevertheless, recent research into the impact of the development of capitalist farming in Britain indicates that women were particularly negatively affected. The sexual division of labour also became much more marked with the capitalisation of agriculture. Historian Pamela Sharpe points out that: "Agrarian capitalism had extremely important effects on the labour force…with larger, capitalised, arable farms [being] likely to employ fewer women and children" (Sharpe 2016, 72–73). In regions of England where enclosure was particularly severe, as in the south-east, the rural population became increasingly dependent upon waged work, which was usually poorly paid, especially for women, in whose case it was very often not even available. According to historian Keith Snell, before 1750 women in the south-east were still engaged in a wide range of agricultural work not in fact markedly different from that of men, but by the mid-nineteenth century women had a very limited role compared to the seventeenth or early eighteenth centuries, with only short-term seasonal work such as weeding in the spring and gleaning after the harvest in August, being unemployed the rest of the time (Snell 1998, 102). Enclosure and the amalgamation of small farms into large ones resulted in a decline in employment for both sexes, although far more so for women; the new agricultural innovations which resulted in larger farms and a transition to arable farming "shed labour, but they shed much higher proportions of female labour than male" (Berg 1994, 137). As a result:

> Severe female unemployment and under-employment by the end of the eighteenth century was a reality. Outside of seasonal work imperatives, married women had few opportunities to work in the market except for commercialised housewifery. Young women either married at young ages, or left rural villages. The route into this for most of them was through domestic service. (Sharpe 2016, 137)

In contrast, based on evidence from estate records, Pamela Sharpe describes how, before the advent of capitalist agriculture, women supported themselves and their families by selling to the estate kitchen their surplus, including "apples to make a syrup of quince…and eggs, pigeons, hops, chickens, mutton, butter, barley, malt, fish, mushrooms, rushlights, cheese, soap and candles" (Sharpe 2016, 84). She points out that "Historians have been slow to recognise the extent to which people relied on the exploitation of their common rights in the past" (84). This lack of recognition of the central importance of access to the commons for small-scale producers to supplement their subsistence production is evident in, for instance, Mingay who, whilst observing that enclosure of the commons often "left the small owner with an acreage too small to be worked without access to a common", nevertheless, considers that "…enclosure…gave a sharp fillip to improved methods of farming" and that the "common-field system was wasteful, expensive, and difficult to work so long as dispersed holdings, common rights grazing, and sharing of the commons remained its basis" (Mingay 1968, 19–22). An even more scathing dismissal of the importance of the commons to people's livelihoods, reflecting the view that, in the interests of modern 'efficient' agriculture, the small 'inefficient' farmer should be eliminated, is expressed by Sir John Sinclair, first President of the Board of Agriculture, which was established in 1793:

> The idea of having lands in common, it has been justly remarked, is to be derived from that barbarous state of society, when men were strangers to any higher occupations than those of hunters or shepherds, or had only just tasted the advantages to be reaped from the cultivation of the earth. (cited in Drummond and Wilbraham 1958, 175)

Right of access to the commons was of particular importance to women whose abilities to support themselves and their families depended upon it. Economic historian, Jane Humphries, challenges the traditional masculinist assumption of most historians that the commons were of relatively minor importance to the livelihoods of the common people, and more particularly to women: "Misled by ahistoric stereotypes of male breadwinners and appurtenant wives and children, economic historians have overlooked the commons' value in facilitating contributions to family subsistence from wives and children" (Humphries 1990, 18). In fact, access to the commons was of particular importance to women, even more than to men, in both supplementing waged labour done by themselves or other members of the family and making an essential contribution to subsistence. Agricultural workers were poorly paid, with women receiving only half to two-thirds of the male wage (38). It was therefore far more cost-effective for them to engage in their own subsistence production using the resources of the commons, such as food-gathering and grazing a cow, than to be dependent upon unreliable low-paid waged work. Access to the commons gave them a degree of independence in providing for themselves and their families, as well as allowing them to combine domestic and production activities, such as baking and brewing, spinning and weaving. In pre-industrial England, the household was the most important unit of production, with "domestic industry centred mainly in the household, and often combined with agriculture, in which women played a central role" (Hill 1994, 15). And, as Humphries points out: "Working at or from home, women could also utilize the labour of their children, and transmit skills" (Humphries 1990, 37). Much of this would have involved the transmission of food production and preparation skills and knowledge, such as dairying, cheese-making, bee-keeping, vegetable growing, baking and brewing.

The cottage industry which enabled women to combine self-employment with household work and the care and nourishment of their children was very much a feature of pre-capitalist production and one that was seriously eroded by the development of capitalism. Unfortunately, this flexibility in women's ability to employ a wide range of skills in their varied work, and their adaptability to different needs and demands in support of the family economy was precisely what made them so readily exploitable by the newly developing capitalist industries: "For merchant manufacturers in search of a plentiful, cheap, and unrestricted labour force, such women provided an ideal and convenient answer – and also a highly exploitable body of labour. Their very flexibility was used against them" (Hill 1994, 46). Access to the commons was also of vital importance for the collecting of fuel with which to cook. With enclosure of the commons this became increasingly difficult; by the end of the eighteenth century, according to social historian, John Burnett, as a result of having to buy fuel rather than gather it from the commons as they had always done, "many labourers went fireless, unable to afford the fuel for ordinary heating or cooking, which had profound effects on the labourer's diet and standard of living generally" (Burnett 1979, 19). Burnett comments that the railing against inferior shop bread and the urging of women to return to home baking by 'parliamentary philanthropists', such as William Cobbett, although well-intentioned, was futile because fuel for cooking was now either no longer available or was beyond the means of the poor.

Maxine Berg states that it is a mistake to impose a modern idea of the nuclear family on the seventeenth and eighteenth centuries, with our "standard image of women working within the context of the family economy being a great distortion for the eighteenth century" (Berg 1994, 157). During this time, women either married late (thirty being the average age of first marriage for women for much of this period) or not at all. In addition, with widows accounting for 12.9% of households, large numbers of unmarried and widowed women were also dependent upon subsistence provided by access to the resources of the commons. Such access was of particular importance to widows; the keeping of a cow, an occupation which required access to the commons for grazing, served a function "equivalent [to] social security" for such women (Humphries 1990, 39). The independence that the use of the commons gave women was resented by capitalist farmers as it meant that these women and their children were engaged in their own activities and production at times when farmers needed labour for harvesting. Because labourers with their own livestock and gardens were "not always at the farmer's 'beck and call', farmers became increasingly intolerant of distractions from waged work, and castigated the commons for obstructing agricultural progress" (29). Although a large proportion of English families were engaged in some waged labour prior to the parliamentary enclosures, the effect of the enclosures was to deprive semi-proletarian families of non-wage sources of subsistence with which they supplemented their waged labour. As a result, they became less self-sufficient and increasingly dependent upon waged labour (Humphries 1990).

Refuting the conservative, anti-Marxist position of J. D. Chamber's classic work '*Enclosure and the Labour Supply in the Industrial Revolution*' (1953) which argues that enclosure had little negative impact on the populace, Humphries states that, if instead:

> proletarianization is seen as a process of gradual elimination of sources of family subsistence other than wages, a causal link between the loss of common rights and wage dependence is re-established ...[and this] analysis must involve the *valuation of common rights and the effects of their alienation*. (Humphries 1990, 19; italics added)

As Humphries emphasises, if we bring women and children back into the picture, we begin to appreciate just how important, indeed essential, the commons were to people's livelihoods: "In the light of the experiences of wives and mothers, historians should revise their view of the value of the commons [so that] hitherto despised activities and denigrated resources [gain] historical significance" (42). Significantly, the view of the commons and peasantry as being backward and unproductive laid the foundations for the colonial expansion and dispossession of Indigenous populations, such as in Australia and the Americas. As Ellen Meikens Woods explains:

> From the earliest years of agrarian capitalism in England, its economic logic became, not just the driving force of the domestic economy, but also an instrument of imperial domination beyond the boundaries of England. The seizure of land that was indeed occupied, and perhaps even cultivated, was justified on the grounds that the occupants had failed to use the land productively and profitably enough, by the standards of English commercial agriculture. (Meikens Wood 2017, 162)

Imperial expansion was also driven by the imperative to provide cheap food for the new urban industrial proletariat. As Patel and Moore explain, "without food surplus, there's no work outside agriculture,..and by 1700 England was doing the two big things that every capitalist power must: increasing the agricultural surplus and expelling labor from the farm" (Patel and Moore 2018, 140–141).

The denigration of the resources and activities that were vital to women's livelihoods is expressed by a contemporary report which disparagingly referred to the products of the commons as "the trifling fruits of overstocked and ill-kempt lands" (cited in Humphries 1990, 21). Such disparagement was both class-based and sexist because it dismissed the vital importance of the commons to the livelihoods of the general populace and to women and their children in particular. The 'despised activities' were those that were essential to the survival of much of the rural populace. These activities, referred to by a contemporary observer as "miserable productions and trifling employments which the heaths in general afford" (cited, 32), involved a vast range of products, including, in addition to fuel for cooking and materials for building houses, "useful dietary supplements from the wild bird and animal life, crab apples and cob nuts from the hedgerows, brambles, whortles and juniper berries from the heaths, and mint, thyme, balm, tansy and other wild herbs…Almost everything in the parish however insignificant could be turned to some good use" (40). As well as using the commons for the gathering of wild food, women also had primary responsibility for cottage gardens, in which they grew food for family consumption as well as selling the surplus. In addition, they often grew cash crops and raised poultry as an extra source of income.

Access to the commons was also necessary for another subsistence activity which was essentially women's work, cow-keeping. This was a particularly valuable activity for the rural household because, apart from providing milk for the family, it also made cheese and butter making possible, all of which added nutritionally to the family diet. This was particularly important for cottagers' children, and where cow-keeping was made impossible by the enclosure of the commons, as it was especially in the south-east of England, a serious deterioration in people's diets occurred. Cow-keeping made an extremely important contribution to both peoples' diets and to household income, with a cow paying for itself within a year. This, however, changed with enclosure, necessitating the purchase of feed which had previously been freely available on common pasture, thus making the keeping of a cow far more expensive and beyond the means of many. Another valuable aspect of cow-keeping was that it enabled the keeping of a family pig, which was fed on whey, the by-product of butter and cheese-making, and on excess garden produce, in addition to foraging for orchard windfalls. Pigs were important to the household diet, especially in winter in the form of bacon used to enrich and flavour many rustic dishes. Humphries recounts that: "Although all family members took pride in their pig, mothers and children were particularly creative in their use of humble resources in its fattening" (1990, 25).

Gleaning, following the harvest, was another area of women's work that made an important contribution to family subsistence. Gleaning, which "from medieval times had (been seen) as beneficial to the farmer as a cleaning operation" (Sharpe 2016, 82), was the one area under capitalist farming from which women benefited; crop specialisation following enclosure, with a greater area given over to a particular crop, meant that, for instance, often enough wheat could be gleaned to provide a household with flour for the year. Nevertheless, by the late eighteenth century gleaning was being increasingly regulated due to farmers viewing it as 'trespassing on private property'. In addition to gleaning, women, especially poor widows with children, sustained themselves by gathering wood, keeping geese or pigs or cows on common land and collecting mushrooms and nuts from fields and woods and berries and other produce from hedgerows. However, "the eighteenth century saw further erosion of these common rights as private property was vigorously defended and gentry farms consolidated" with women being increasingly charged with the illegal gathering of wood (Sharpe 2016, 85).

Feminist sociologist, Carolyn Sachs explains that the privatisation of the commons, which has become a global phenomenon due to colonisation and the increasing privatisation of land by capitalist development of agriculture, has been justified by the "tragedy of the commons", a theory put forward by Garrett Hardin in 1968. This theory, now widely accepted as fact, posits that human beings will inevitably overexploit the resources of the commons, leading to their inevitable collapse. Sachs points out that amongst the numerous problems with this theory, including its Eurocentric perspective on the concept of property, is that it assumes an individualistic approach to the commons in which people seek maximum short-term gains. Traditionally, however, users of the commons, often women, worked in co-operation with each other, and had strong incentives not to overexploit the resources, as their survival depended upon them. The takeover of the commons for capitalist exploitation is now occurring throughout much of the Global South, including land grabbing by international agribusiness interests, depriving rural and peasant women of food and fuel with which to sustain their families. As Sachs states, the 'real tragedy' is the global decline in access to the commons as a result of modern capitalist development, meaning "that women are hungrier, more isolated, and less able to provide for their families" (Sachs 1996, 48).

As access to the resources of the commons was especially important to women and to the support of their children, not surprisingly it was women who were often at the forefront of resistance to enclosure. Maxine Berg points out that there is evidence that women, being "an important section of these rural communities frequently played a leading part in local custom and protest, [and that] they led food riots ...[and] mobbed poor-law officials" (Berg 1994, 164). British historian E. P. Thompson elaborates further:

> Women's prominence in food riots...was due to their role in the economy, a role which gave them 'authority and self-confidence.' The women had charge of marketing for provisions, of baking, brewing, and feeding the household. They knew about price and quality, and through their neighbourhood connections they also took the initiative in bonding households together in times of protest. (cited in Berg 1994)

It is very interesting and important to note that in recent years, some historians—for instance Peter Linebaugh and Silvia Federici—have analysed the connection between the development of capitalist agriculture and the witch-hunts which occurred in Europe during the sixteenth and seventeenth centuries. As described above, rural women were particularly adversely affected by the enclosure of the commons. This was because of their dependence on them as a source of food, fodder and fuel for their subsistence and economic independence, and once deprived of these they were reduced to poverty and homelessness. Older women and widows were especially negatively affected; prior to the enclosures, in England widows and poor people in general had been provided for by the manorial system. Under this system, which had existed since the Magna Carta and the Charter of the Forest of 1215, widows had been given a guaranteed right to food, sustenance and wood for fuel (Federici 2018, 18). As a result of the enclosures of the commons, these rights no longer existed, and dispossessed rural women, especially older women and widows were reduced to wandering the countryside and begging. This was seen as disruptive to social order, with these women often being accused of being witches, especially when they "resisted their impoverishment and social exclusion", 'making nuisances' of themselves by begging, and participating in protests by pulling up fences that now enclosed the commons (2018, 19). The 'anti-social' behaviour of these dispossessed and often justifiably angry women resulted in many being accused of witchcraft and blamed for the failure of harvests, the death of farm animals and even children, leading to accusations of consorting with the devil, and subsequent torture and execution. This happened on a horrific scale throughout much of Britain and much of Europe during the sixteenth and seventeenth centuries when hundreds of thousands of women were executed as witches.[1]

The displacement of the rural populace caused by enclosure and the accompanying rapid urbanisation of England as a consequence of the Industrial Revolution had major and far-reaching consequences with regard to food. The population became more and more divorced from the land which had previously supported them. In the:

> new urban environment, traditional rural skills, of which domestic baking and brewing were economically the most important, gradually decayed and ultimately disappeared over wide areas of the country. In earlier times the housewife brewed as naturally as she baked, and prided herself on her skill in producing different varieties, strengths and flavours of beers… Brewing, even more than baking, presupposed a standard of living above mere subsistence. However, by 1815, regular domestic brewing had ceased over most of the country. (Berg 1994, 17–19)

The industrialisation of bread and beer resulted, not only in loss of skills and income for women, but also in an inferior, often adulterated product. The sad irony is, that in recent times, as a reaction against the often bland, homogenised taste of industrial bread and beer, men have enthusiastically taken to artisan bread and beer making, with, in typical masculinist tradition, no acknowledgement or recognition that these are in fact traditionally women's skills going back millennia.[2] Interestingly, and significantly, in remarking on the important role of women in providing food for the family, including the sale of surplus—as well as in the production of other goods, such as wool, flax or hemp, to clothe their families—and how this was lost as a result of enclosure and industrialisation in England, historian Bridget Hill comments on the fact that, in contrast, rural French women, even today, still take pride in their contribution to family subsistence:

> In France to the present day, wives of small peasant farmers refer always to *mes vaches* and *mes poulets*. They are their responsibility and they take an understandable pride in them. When they sell the milk, butter, cheese

> eggs or poultry, the money goes into their own pockets and provides them with a measure of financial independence. (Hill 1994, 45)

However, for English women who had once taken a similar pride in such subsistence activities, this was no longer possible due to the process of enclosure. And, crucially, what was also lost with the loss of access to the resources that formed the material base of this cuisine was *the knowledge that went with it*—that is, the knowledge that had been accumulated over many centuries about cheese-making, curing, brewing, preserving, cultivating, and gathering. Such knowledge had been passed on from one generation of women to the next, from mother and grandmother to daughter and granddaughter.

The negative impact that the elimination of the peasantry had on rural English cooking was commented on by Eliza Acton who, at the time, lamented the loss of the "art of preparing good, wholesome, palatable soups, without great expense, which is so well understood in France, and in other countries where they form part of the daily food of all classes of the people" (cited in Spencer 2002, 246). As English food historian Dorothy Hartley vividly expressed it: "With the Industrial Revolution the people are driven off the land; they are cut off from their natural food supply, and are compelled for the first time to *buy* food. The cruelty of the Industrial Revolution was that it made money a necessity of life…(and caused) *a dislocation of the food supply*" (Hartley 1954, 579; original italics). Once this dislocation had happened it was permanent; the displaced rural population who moved into the industrial towns in search of work could not return to the country because agricultural productivity requires continuity and constant tending. Therefore, for "the unfortunate victims of the industrial townships, there was no 'return home'. Once away from their basic proportion of arable land, they were dependent, for the first time in history" (Hartley 1954, 579).

> The growth of a market economy had …undermined local self-sufficiency and enmeshed the village in a network of cash sales…What this change had wrought was *to erase the English rural cuisine*: there were no ingredients for the soup or pottage flavoured with wild greens… (Spencer 2002, 246; italics added)

The dislocation from the land and dependency on waged labour is reflected in the fact that "by 1850 the individual household had become dependent on professional food suppliers" (Burnett 1979, 21). The dispossession consequent upon the enclosures and the subsequent forced dependency on low-waged work resulted in an appalling diet of 'bread, jam, tea and sugar'. Reporting on the health of women labourers, one contemporary doctor noted that four out of five women who sought treatment suffered from ailments that were a result of their food being "insufficient in quantity and not good enough in quality" (Burnett 1979, 43). In addition to the serious decline in health was the loss of food knowledge; the loss within a few years of "those habitual daily recipes which had been a central part of their life", with the result that "the recipes that had once been passed from generation to generation now vanished" (Spencer 2002, 210). This experience was unique to Britain where the Industrial Revolution occurred much earlier and on a much larger scale than elsewhere. This meant that, with the exception of some more remote parts of Britain that escaped the impact of industrialisation, much of the rural cooking tradition of the British Isles was lost.

The poor living conditions and long working hours of the urban proletariat during the period of industrialisation placed serious restraints upon cooking and the preparation of meals, and that "to be concerned with meals reflected leisure, wealth and the availability of domestic labour" (Oddy 2003, 2). Speaking of the decline in the standard of nutrition of Scottish workers in the jute industry, one commentator observed that: "When the mother is at work there is not time to prepare porridge or broth...usually breakfast and dinner become bread and butter meals" (cited in Mintz 1985, 128). The virtual disappearance of traditional homemade soups, broths and pottages was a result of such restraints on cooking; consequently "the fact that people were satisfied to eat bread with cold meat or cheese and beer as regular everyday fare in much of England may well have been the principal factor limiting the culinary arts in Britain" (Oddy 2003, 4). However, rather than people being 'satisfied' with such a diet, it would be more accurate to say people were *forced* by their changed circumstances into such a limited diet because industrialisation and urbanisation had destroyed much of the pre-existing food and culinary knowledge that had been developed over many centuries by generations of women in a previously self-sustaining rural economy. The increasing urbanisation which reduced a connection to the land, as well as the decline in domestic food production as a consequence of changes in land organisation, resulted in a reduction or loss of traditional supplies of milk, potatoes, vegetables, pigs and chickens, a loss which remained a major source of grievance throughout the nineteenth century (2003).

The Scientific Revolution and the Development of Industrial Agriculture as a By-Product of War

The Scientific Revolution of the sixteenth and seventeenth centuries slightly preceded the Industrial Revolution and then developed in tandem with it, each supporting and reinforcing the other. The Scientific Revolution represented a fundamental shift in the Western world's view of nature, from one in which the cosmos was seen as an organic unity in which "all parts of the cosmos were connected and interrelated in a living unity…(with) all parts of nature mutually interdependent….,each reflect(ing) changes in the rest of the cosmos" (Merchant 1989, 103) to a mechanistic view of nature as something to be controlled and dominated. This change in attitude to nature laid the foundations for the later development of industrial food production, as well as the wholesale exploitation of the earth's natural resources, starting with the Industrial Revolution and the beginning of the coal industry, which has over the last couple of centuries led to the environmental crisis that we are now experiencing, especially with regard to climate change.

> The Scientific Revolution of the sixteenth and seventeenth centuries has been treated by most historians as a period of intellectual enlightenment in which a new science of mechanics and a mechanical world view laid the foundation for modern scientific, technological and social progress. But, in the face of the current crisis over the depletion of natural resources, Western society is once more beginning to appreciate the environmental values of the pre-mechanical 'world we have lost'. (Merchant 1989, 99)

The most significant figure in developing the mechanistic view of nature as something to be dominated and subdued was Francis Bacon (1561–1626) who is regarded as the father of Western science. Bacon considered nature a chaotic and unruly force that needed to be subdued and brought under control; his thinking was strongly influenced by the witch-hunts that were taking place during the early seventeenth century in much of Europe, including his native Britain (168). Indeed, in his writings, he likened the control of nature to females persecuted as witches; nature needed to be interrogated and tortured using mechanic devices such as those used in the inquisition of witches. The language used was akin to that of rape:

> The new man of science must not think that the 'inquisition of nature is in any part interdicted or forbidden'. Nature must be 'bound into service' and made a 'slave', put 'in constraint' and 'moulded' by the mechanical arts... *Neither ought a man to make scruple of entering and penetrating into these holes and corners, when the inquisition of truth is his whole object.* (Merchant 1989, 68–169; italics added)

According to Bacon, the most noble and ambitious of human ambitions was "to endeavour to establish and extend the power and dominion of the human race itself over the universe", so that "the human race (could) recover that right over nature which belongs to it by divine bequest" (1989, 172). Bacon was not alone in the quest to manipulate and control nature; a number of other philosophers, scientists and mathematicians of the sixteenth and seventeenth centuries promoted a similar mechanistic view of nature, including Robert Boyle, Hobbes, Mersenne, Gassendi and perhaps most notably French philosopher and mathematician Rene Descartes, who in his '*Discourse on Method*' advocated "render(ing) ourselves the masters and possessors of nature" (1989, 188). This mechanistic approach to nature, in which the world is viewed not as an integrated organic whole, but as something to be dissected and remodelled to suit the interests of capital has formed the basis of modern food production and agriculture.

Although the foundations of this approach were developed two or three centuries ago, it has, especially since the Second World War become the main mode of food production with all the attendant environmental, social and health problems. The development of industrialised agriculture was facilitated and driven by the development of the fossil fuel industry and as a by-product of the First and Second World Wars. Indeed, one of the most sinister aspects of industrial agriculture is that common pesticides and herbicides were derived from deadly chemicals used in war. Products of war designed to kill people were later adapted to wage war on nature, with "the large manufacturing facilities producing wartime nitrates (for bombs) and toxic chemicals (for poison gas) being refitted to produce fertilizers and pesticides" (Holt-Giménez 2017, 45). One notorious example is that of the world's best known and most widely used herbicide, Roundup, which was derived from the infamous

Agent Orange, developed by the chemical company Monsanto, and which contains deadly dioxin, being used by the United States on a huge scale as a defoliant during the Vietnam War (Robin 2010, 3). This has resulted in a toxic legacy for the Vietnamese population (as well as US and 'allied' soldiers) causing cancers and birth defects to this day.

Although Monsanto (bought in 2018 by German chemical company Bayer) describes itself as an agricultural company, its principal interest is in making and marketing chemicals, including the most widely used herbicide in the world, glyphosate-based Roundup Ready. Another infamous product associated with Monsanto is its role in developing genetically modified (GMOs) seeds engineered to withstand being sprayed with Roundup. In her book on the history of Monsanto's manipulation and corruption of the science and denial of the harms of glyphosate, American investigative journalist, Carey Gillam describes how the development of Monsanto's genetically modified crops in the 1990s arose out of its determination to control and dominate the global herbicide industry. In order to retain its dominant position, Monsanto's scientists developed genetically altered soybeans, as well as other crops, including maize, cotton and sugar beets. As a result, these 'Roundup Ready' crops marked the dawning of a new era for agriculture that would quickly and drastically change the face of food production" (Gillam 2017, 40). Of the many myths and lies promoted by Monsanto, perhaps the most cynical and dishonest is its claim that GMOs are necessary to feed the world. Far from this being so, GMOs epitomise the agro-industrial corporate complex which is both a major threat to food security and food sovereignty, an issue that I address in more detail in the following chapter. In short, GM crops and the industrial monocrops that they represent are the ultimate antithesis of the environmentally sustainable agroecological methods of food production practised by the majority of the world's small-scale farmers, most of whom are women—a subject detailed in Chapter 8.

Industrial agriculture, with its dependence on high chemical inputs to be productive, and epitomised by GMOs, is the product of the reductionist, masculinist, mechanical approach to nature initiated by the 'father of modern science', Francis Bacon in the late sixteenth century. Furthermore, with its total reliance on fossil fuels, from the use of artificial fertilisers, factory farming of animals, deforestation, processing and packaging and transport the industrial food system is a major threat to climate stability, contributing between 44 and 57% of global greenhouse gas emissions (GRAIN 2016, 2–3).

Adulteration of Food

One of the most pernicious effects of the industrialisation of food has been the adulteration of food. In his study of the changes in the British diet from the 1890s to the 1990s, food and nutrition historian, Derek Oddy notes that before the Industrial Revolution the British diet was characterised by local domestic production and seasonal availability, with traditional methods such as smoking and salting being used to preserve foods. However, as much of Britain became rapidly industrialised and urbanised during the eighteenth and nineteenth centuries, such local and seasonal domestic production drastically declined, with only a few areas such as the remote uplands still retaining the vestiges of a rural peasant-based economy. "Rural society, as a self-sustaining system, was in a terminal phase" (Oddy 2003, 1); the urban and semi-rural proletariat were now dependent upon waged labour and the purchase of increasingly industrialised, imported and often adulterated foods.

Although manufacturers of food have long sought to cut costs and increase profits by adulterating food, it was during the Industrial Revolution, when society became highly urbanised, and with people no longer being connected to the source of their food, that food adulteration became a serious problem. Previously, when England was still largely an agricultural society, "each village – and each household within the village – was to a large extent self-sufficient, and every housewife was her own baker, butcher and brewer" (Burnett 1979, 100). However, with rapid industrialisation and resulting urbanisation, the population became increasingly dependent on purchased, processed food, so that by the early 1800s fraudulent adulteration of food proliferated, with "far-reaching social, economic and medical consequences" (1979, 99). Commercial bread was whitened with alum, a chemical used since medieval times in the textile industry for fixing dyes (Wilson 2009, 77); a variety of additives, some of them poisonous, were used in commercial ale and beer as a substitute for hops and malt (Burnett 1979, 102); and children's sweets were coloured with toxic dyes, such as lead and mercury and copper-based dyes (Wilson 2009, 113).

In England bread had been baked on a relatively small commercial scale for centuries. Its quality had been ensured since 1266 when King Henry III promulgated the Assize of Bread and Ale, which "regulated the weight, price and composition to a remarkably precise degree" (Wilson 2009, 66). However, this law was abolished more than five centuries later in 1822, opening the floodgates to the fraudulent adulteration of food, particularly of bread and beer. This happened, paradoxically, in the wake of the publication of expatriate German chemist Frederick Accum's *Treatise on the Adulteration of Food, and Culinary Poisons* which exposed the shocking extent of food adulteration in England in the early nineteenth century. Because of the British government's belief in unregulated free trade and *laissez-faire* economics, any controls on the quality of food were abolished, with the result that food adulteration proliferated more than ever. This affected the urban poor much more than the well-off, who, as always, were able to procure their food from reliable sources, as well as having the means to employ their own cooks, unlike the poor who had long since lost the ability to grow and cook their own food for the reasons explained earlier in this chapter. The *laissez-faire* approach of the British government stood in stark contrast to the policies of several other European countries, such as France where there was a board of health "whose job it was to watch over anything that might harm public health, including adulterated food" (111).

Additives to food increased from the mid-nineteenth century with the development of large-scale manufacture of processed food. Although there are now more regulations against the fraudulent adulteration of food, so that commercial food is generally safer now than in Victorian times, fraudulent tampering with food is by no means a thing of the past, with "food fraud (being) no more a relic of history than greed and deceit are" (Wilson 2009, 270). Rapid industrialisation and urbanisation in China over the past couple of decades has led to rampant adulteration of food. The most infamous was the sale of baby milk powder bulked out with melamine, which led to the death of several babies in the early 2000s. Although China's small dairy farmers were not responsible for this contamination, their livelihoods were destroyed because of the lack of demand for domestic production as a result of the contamination. However, China's big dairy companies profited by establishing mega dairy farms both in China and abroad to produce powdered milk (GRAIN 2016, 176). This and other food scandals have led to widespread distrust on the part of Chinese consumers who now, for instance, buy up huge stocks of infant milk formula from countries considered clean and safe, such as Australia and New Zealand. The good reputation of food from these countries has in turn fostered further food scandals, such as the selling of inferior products, including fruit, meat and wine, being passed off as the genuine products ("Spot the fake" *The Age*, 8 August 15).

Perhaps the most infamous case of food contamination in recent times was that of bovine spongiform epidemic (BSE) in Britain, in which, as a consequence of neoliberal economic policy, cattle-feed was adulterated with the rendered carcasses of sheep deliberately processed at a lower heat in order to reduce costs. Apart from the obscenity of feeding herbivores with processed animal remains, the practice resulted in deaths from incurable Creuzfeldt Jacob Disease (CJD) brain disease in people who had eaten the contaminated beef.

Since the Second World War, adulteration of food in the form of food additives has burgeoned on a massive scale. The food industry has more than 4,500 different flavouring compounds available to "create the impression of fruit or vegetables or other expensive ingredients without having to bother to pay for the real thing" (Schlosser 2001, 200). In his analysis of the American fast-food industry, Eric Schlosser states that in the United States approximately ten thousand new processed food products are introduced every year, with most of them requiring a battery of artificial flavour additives (124). As British food journalist Joanna Blythman explains, with factory-made food now being the dominant way of eating for most people, it is increasingly difficult to avoid the 6,000 food additives existing "that are routinely employed behind the scenes of contemporary food manufacture (Blythman 2015, 3). By the early 2000s it was estimated that consumers in industrialised countries were eating between six to seven kilogrammes of food additives per year (Millstone and Lang 2003, 90).

Industrialised Animal Production

Probably the worst and most cruel manifestation of industrialised food systems is the industrialisation of animal agriculture, in which animals are treated merely as commodities, not sentient beings, with maximisation of profit being the primary concern, regardless of the suffering of the animals. As academic agronomist Fred Magdoff graphically puts it: "Torture a single chicken and you risk arrest. Abuse hundreds and thousands of chickens for their entire lives? That's agribusiness" (cited in Holt-Giménez 2017, 81). The horrors of industrial animal production (it cannot be called farming in the true sense) are well known and well documented: tens of thousands of chickens crammed into indoor spaces with scarcely enough room to move, and deprived of their natural needs, including sun and the ability to run and forage; similarly with pigs in huge numbers crowded together and raised in cramped stalls, fed antibiotics to deal with the diseases inevitable in such crowded conditions, and hormones to promote fast growth; dairy cattle injected with bovine growth hormone to stimulate milk production causing severe mastitis and shortened lives in cows; beef cattle confined in their thousands, standing ankle deep in their own manure in feedlots…the list goes on. The push by governments and agribusiness to produce cheap meat that has taken place since the Second World War has led not only to unspeakable cruelty to billions of animals, but it also is the cause of health problems in humans as a consequence of overconsumption of meat. Furthermore, industrial meat production has become a major contributor to climate change, largely as a result of deforestation in countries such as Brazil for the industrial-scale production of soy and maize (mostly genetically modified) grown for export as feed for industrially 'farmed' cattle, pigs and chickens.

An analysis of the development and effects of industrialisation of food is important because, although this model now exists in much of the developed world, it was Britain, and more specifically England, due to its early industrialisation, that was the first country to experience this transformation of the food system from one of peasant self-sufficiency to one of dependence on agribusiness and the corporate food industry. Speaking of the alienation of most Britons from food production, their ignorance and distrust of raw ingredients and their ever-increasing reliance on industrially processed food, Joanna Blythman comments that: "In a country that industrialized early, where a mutual sea of incomprehension lies between the urban and the rural, relatively few Britons have any intuitive feel for food" (2006, 195). Unfortunately it is this model that is increasingly becoming the norm throughout the entire world, bringing with it all the attendant environmental, health and social problems. As food policy expert, Tim Lang puts it: "Britain, which due to its early industrialization and colonial empire had long ago stopped feeding itself, is perhaps a forerunner of a pattern now emerging elsewhere" (Lang 1999).

Tragically it is this model, what ecofeminist Maria Mies calls capitalist accumulation, that is being adopted by and imposed on much of the world as new frontiers open up and large-scale, industrial, capitalist agriculture and food production move in. This is replicating the disasters of the industrialised West, pushed by agribusiness and corporate governments, undermining small-scale, traditional, biodiverse systems of food production in these countries and the culture that is both a product of them and that supports them. Furthermore, crucially, industrialised food production undermines/destroys food knowledge and skills, both in the growing and preparation of food, traditionally areas of female knowledge and expertise, as I have described in this chapter. Remarking on the contrast between industrialised agriculture in Britain and a still largely self-sufficient rural economy in France in the late 1940s, English farmer Philip Oyler lamented the loss in England of the small-scale, highly diverse and productive farms that he found in the Dordorgne Valley of south-west France. He also foresaw the disastrous consequences of industrial agriculture which had grown out of the British scientific and industrial revolutions: "The Industrial Revolution destroyed that precious heritage for us. The link with the past has not been twisted. It has been broken, and we shall have to relearn our lessons, for scientific agriculture is only leading mankind into creating deserts and world-wide famine" (Oyler 1951, 34). Indeed, agriculture has become far more industrialised and intensive during the decades since this was written. Industrialised agriculture is now the dominant agriculture throughout the Western and developed world, even in a food-conscious society like France—which is, in fact, the fourth largest user of pesticides in the world (Robin 2014, 53)—as well as increasingly in the Global South.

As I have outlined in this chapter, the development of capitalist industrialised food systems is the outcome of a Western masculinist mechanistic approach to food production. Instead of viewing access to affordable nutritious food as an inherent human right, it has become a commodity whose main aim is to maximise profits, with massively negative consequences for the environment, human health and animal welfare. This exploitative approach to food production stands in stark contrast to holistic agroecological food systems traditionally practised for millennia by peasant farmers throughout the world, and which I describe in Chapters 7 and 8. These integrated and sustainable food systems are still the norm throughout the non-industrialised Global South, and where a majority of small-scale farmers are women whose ability to produce food is increasingly under threat from the incursion of industrialised farming systems.

The consequences of the food industry include serious environmental degradation, pollution and climate-changing greenhouse gases, inhumane factory farming of animals and an epidemic of obesity and related health problems. The industrial model of agriculture which is now the dominant one in Western industrialised countries, is not only destructive of the environment and people's health, but is also, with its profligate use of resources and environmentally damaging practices, fundamentally unsustainable. Nowhere has the industrialisation of food been more apparent than in the United States, with the health and environmental consequences of such an industrialised way of eating having become increasingly apparent over the last few decades. This is summed up by American nutrition academic, Marion Nestle:

> The US food industry is the remarkable result of twentieth-century trends that led from small farms to giant corporations, from a society that cooked at home to one that buys nearly half its meals prepared and consumed elsewhere, and from a diet based on 'whole' foods grown locally to one based largely on foods that have been processed in some way and transported long distances. (Nestle 2002, 11)

It is this model that is increasingly being imposed on and adopted by the rest of the world, bringing with it the many harms associated with the industrialised food system. These issues are addressed in more detail in the following chapter.

Notes

1. The witch craze occurred, not only in England, but also throughout much of Europe during the sixteenth and seventeenth centuries, especially in southern Germany, Switzerland and parts of France and Italy. Although the causes varied from region to region, those accused of witchcraft were overwhelmingly women, and usually, although not exclusively, older single or widowed women. Accusations of witchcraft not only originated with the general population, but were encouraged by the Church, especially after the publication in 1486 of the infamous *Malleus Maleficarum*, by the Dominican priests Kramer and Sprenger. For more on this see 'Witch Craze' by Lyndal Roper and 'Gyn/Ecology: the Metaethics of Radical Feminism' by Mary Daly, 'Witches, Midwives and Nurses: A History of Women Healers' by Barbara Ehrenreich and Deidre English]. Significantly, Silvia Federici notes that accusations of witchcraft against rural women now exist/have reappeared in parts of India and Africa, driven largely by globalisation and economic liberalisation.
2. Food writer, Simran Sethi describes how in the Middle Ages women were the main beer makers and were known as 'alewives' or 'brewsters'. "This work gave women an income, social standing and, if married, a degree of independence from their husbands" (*Bread, Wine, Chocolate: The Slow Loss of Foods We Love*, 2015, 224). However, given that people drank one

litre of beer a day, it was a very lucrative enterprise, which resulted in men taking it over, and with laws limiting the amount of barley women could buy (224).

References

Berg, Maxine. 1994. *The Age of Manufacuturers 1700 to 1800: Industry, Innovation and Work in Britain*. London and New York: Routledge.

Blythman, Joanna. 2006. *Bad Food Britain: How a Nation Ruined Its Appetite*. London: Fourth Estate.

Blythman, Joanna. 2015. *Swallow This: Serving Up the Food Industry's Darkest Secrets*. London: Fourth Estate.

Burnett, John. 1979. *Plenty and Want: A Social History of Diet in England from 1815 to the Present Day*. London: Scolar Press.

Drummond, J.C., and Anne Wilbraham. 1958. *The Englishman's Food: A History of Five Centuries of English Diet*. London: Jonathan Cape.

Federici, Silvia. 2018. *Witches, Witch-Hunting and Women*. Oakland, CA: PM Press.

Gillam, Carey. 2017. *Whitewash: The Story of a Weed Killer, Cancer and the Corruption of Science*. Washington, DC: Island Press.

GRAIN. 2016. *The Great Climate Robbery: How the Food System Drives Climate Change and What We Can Do About It*. New Delhi: GRAIN/Daraja Press and Tulila Books; Melbourne: Spinifex Press.

Handy, Jim, and Carla Fehr. 2011. "Drawing Forth the Force that Slumbered in Peasant's Arms: The 'Economist', High Agriculture and Selling Capitalism." In *Food Sovereignty: Reconnecting Food, Nature and Community*, edited by Hannah Wittman et al., 45–61. Cape Town, Dakar, Nairobi, and Oxford: Pambazuka Press.

Hartley, Dorothy. 1954. *Food in England*. London and Sydney: Futura.

Hill, Bridget. 1994. *Women, Work and Sexual Politics in Eighteenth Century England*. London: UCL Press.

Holt-Giménez, Eric. 2017. *A Foodie's Guide to Capitalism: Understanding the Political Economy of What We Eat*. New York: Monthly Review Press.

Humphries, Jane. 1990. "Enclosures, Common Rights, and Women: The Proleterianization of Families in the Late Eighteenth And Early Nineteenth Centuries." *The Journal of Economic History* (1): 17–42.

Lang, Tim. 1999. *Diet, Health and Globalization: Five Key Questions*. Paper presented at the International and Public Health Symposium, 'Feeding the World', University of Surrey, June.

Meikens Wood, Ellen 2017. *The Origin of Capitalism: A Longer View*. London and New York: Verso.

Merchant, Carolyn. 1989. *The Death of Nature: Women, Ecology and the Scientific Revolution*. New York: HarperCollins.

Millstone, Erik, and Tim Lang. 2003. *The Atlas of Food: Who Eats What, Where and Why*. London: Earthscan.

Mingay, G.E. 1968. *Enclosure and the Small Farmer in the Age of the Industrial Revolution*. London and Basingstoke: Macmillan.

Mintz, Sidney W. 1985. *Sweetness and Power: The Place of Sugar in Modern History*. New York: Viking.

Nestle, Marion. 2002. *Food Politics: How the Food Industry Influences Nutrition and Health*. Berkeley, Los Angeles, and London: University of California Press.

Oddy, Derek J. 2003. *From Plain Fare to Fusion Food: British Diet from the 1890s to the 1990s*. Woodbridge, Suffolk, UK and Rochester, NY: Boydell Press.

Oyler, Philip. 1951. *The Generous Earth*. London: Penguin.

Patel, Raj, and Jason M. Moore. 2018. *A History of the World in Seven Cheap Things: A Guide to Capitalism, Nature and the Future of the Planet*. Melbourne: Black Ink.

Robin, Marie-Monique 2010. *The World According to Monsanto: Pollution, Politics and Power*. Translated from the French by George Holoch. North Melbourne, Australia: Spinifex Press.

Robin, Marie-Monique. 2014. *Our Daily Poison: From Pesticides to Packaging, How Chemicals have Contaminated the Food Chain and are Making Us Sick*. New York and London: The New Press.

Sachs, Carolyn. 1996. *Gendered Fields: Rural Women, Agriculture and Environment*. Boulder, CO and Oxford, UK: Westview Press.

Schlosser, Eric. 2001. *Fast Food Nation: What the All-American Meal Is Doing to the World*. London: Allen Lane; New York: Penguin Press.

Sharpe, Pamela. 2016. *Adapting to Capitalism: Working Women in the English Economy, 1700–1850*. Basingstoke, UK: Palgrave Macmillan.

Snell, K. D. M. 1998. "Agricultural Seasonal Unemployment, the Standard of Living and Women's Work (1690–1860)." In *Women's Work: The English Experience 1650–1914*, edited by Pamela Sharpe, 73–120. London and New York: Bloomsbury.

Spencer, Colin. 2002. *British Food: An Extraordinary One Thousand Years of History*. London and New York: Routledge.

Sutton, David E. 2001. *Remembrance of Repasts: An Anthropology of Food and Memory*. Oxford and New York: Berg.

Wilson, Bee. 2009. *Swindled: From Poison Sweets to Counterfeit Coffee—The Dark History of the Food Cheats*. London, UK: John Murray Press.

CHAPTER 7

Threats and Solutions to Biodiversity, Cultural Culinary Diversity and Food Sovereignty

The displacement from the countryside and the accompanying loss of culinary skills and knowledge that occurred during the Industrial Revolution is now increasingly occurring in countries of the Global South. As a consequence: "Similar conditions accompany displacement from the countryside in Third World countries today, as severe as those of early industrial England" (Friedmann 1999, 50). Indeed, in many instances, this displacement is even greater than that in early industrial England, especially in recent decades as a result of a number of additional factors, including land grabbing, climate change, and neoliberal policies of free trade promoted by the World Trade Organisation (WTO). Colonialism in countries of the Global South has historically involved the cultivation of export crops to the colonising nations, a process that has often had devastating consequences on production of food for local consumption. This process accelerated during the latter part of the twentieth century with the imposition of Structural Adjustment Programmes (SAP) by the International Monetary Fund on developing countries crippled by foreign debt. One of the principal ways in which SAP are applied in order to service debt is to encourage exports, including agricultural exports. This results in the displacement of indigenous subsistence food crops by cash crops grown for export, with dire consequences for local producers and consumers, especially small-scale subsistence farmers and the urban poor. These destructive effects are further exacerbated by free trade agreements

V. A. Swinbank, *Women's Food Matters*,
https://doi.org/10.1007/978-3-030-70396-7_7

that benefit, at the expense of small farmers, large-scale industrialised farming controlled by agribusiness and transnational corporations (TNC), all promoted by neoliberal capitalism since the 1980s.

The Neoliberal Food Regime

Mexico is a glaring example of the destructive consequences of a neoliberal capitalist approach to food issues. In Mexico, as a result of the North American Free Trade Agreement (NAFTA), signed in 1994 and SAP, local food production has increasingly been replaced by crops grown for export to the industrialised North, a process which is causing the "old links between local agriculture and local cuisines to be replaced by a new dependence on distant buyers and sellers" (Friedmann 1999, 50). NAFTA resulted not only in the loss of hundreds of thousands of jobs in the United States, but also the displacement of two million Mexican peasant farmers, who "unable to compete with US corn, have been forced off the land" (Hauter 2012, 27). A graphic example of this displacement is that of Mexican women employed by agribusiness companies picking and packing tomatoes for the United States and Canadian markets, which demand a year-round supply, regardless of the season. As Harriet Friedmann explains, most of the workers in this industry are women from poor peasant backgrounds who cannot afford to buy the food they process for export, nor are they any longer able to grow any of their own food. As a result, they are forced to spend their meagre wages on buying cheap imported US corn that is also likely to be genetically engineered, and which is different in taste and texture from the traditional corn used to make the tortillas that form the basis of the Mexican diet. No longer having the time to produce the labour-intensive and nutritious traditional corn tortillas, these women are forced to resort to industrial tortillas produced by US companies such as Taco Bell. Indeed, in a tragic paradox, Taco Bell opened branches in Mexico serving "industrial tortillas to Mexicans uprooted from their village and urban networks" (Friedmann 1999, 48).[1] As the commodification of food is intensified nearly everywhere, more and more people in Mexico and in other countries around the world, many of them from peasant communities, work to produce food that they may not ever eat themselves. Instead, they are supplying world markets and often eat what comes back to them from world markets (48).

Deborah Barndt points out that the dilemma for low-paid women having to resort to industrialised processed food to feed their families due to the lack of time and resources is also one shared by many women in industrialised developed countries. For example, Canadian supermarket cashiers and fast-food workers, most of whom are women, are forced to adapt to the flexible labour strategies of globalised business interests, which entails lack of job security and irregular working hours, all of which affects "family routines, interactions and, especially, eating practices" (Barndt 1999, 3). As US-based multinational food corporations, such as McDonald's and Taco Bell, seek new markets in the developing world, they impose on other cultures an American capitalist model of standardisation and homogenisation. This capitalist model involves low wages, low environmental standards and promotes a system of industrialised food production that presents a serious threat to both cultural diversity and biodiversity, as well as people's health. It also stands in direct opposition to a "truly cosmopolitan world culture [which] would consist of many local cultures existing together, mutually and respectfully exchanging and learning from one another" (56).

From being the historical centre for the development of maize thousands of years ago with a vast variety of landraces, and upon which the richly diverse regional healthy and sustainable culinary cultures were based, Mexico is now a tragic showcase of genetic erosion caused by the flooding of the country with hybrid and genetically modified maize from the United States. Furthermore, the trade liberalisation arising from free trade agreements, namely NAFTA, has resulted in the dumping of cheap mass-produced industrially processed foods on the Mexican market, leading to an epidemic of obesity in the Mexican population. Additionally, the flooding of Mexico with cheap subsidised corn from the United States has made maize growing less viable for many small-scale Mexican farmers, who are then driven to attempt to 'illegally' migrate in order to find work north of the border in the United States.

Free trade agreements, much of it concerned with agriculture, inevitably have a harmful effect on small-scale farmers, the majority of whom are women in the Global South. In Africa, where some countries have signed free trade agreements with the European Union and the United States, little has been gained on the African side, with for instance exports from Africa declining from US$68.2 billion to US$23.2 billion in 2014. (GRAIN 2020). A recently ratified African Continental Free Trade Agreement, an initiative of the African Union is intended to "boost trade between African countries, and ultimately encourage and reinforce regional integration" (GRAIN 2020). Although on the surface this may sound positive, in fact such an agreement will likely follow a neoliberal agenda of opening the floodgates to promoting corporate industrial agriculture and the privatisation of seeds, and potentially opening "the door to full-fledged patenting of transgenic seeds at a later stage" (4). This will have disastrous consequences for rural women who, as seed custodians, play a critical and central role in African seed systems, managing "the preservation, diversity, selection and storage of seed in most if not all African communities" (2020, 4). Even though these activities are essential to rural household well-being, they are not considered as 'economically active employment' in national accounts. (Waring 2004) Crucially, however, "if women cannot continue to play their role in seed saving, their socio-economic position, their leverage in decision making processes and ultimately their power as women is compromised and eroded" (GRAIN 2020, 5).

Threats to Women's Role as Custodians of Biodiversity

Virginia Nazarea asserts that, especially in subsistence societies, such as those she has studied in the Philippines, it is women who hold the key to the preservation of both biodiversity and the cultural memory that supports it: "It appears that resistant memories, empathy towards biodiversity, and irreverence to the treadmill of monoculture are exhibited to a greater degree by subsistence farmers than commercial farmers, and *more by women than by men*" (Nazarea 2005, 115; *italics added*). This is because of women's central role in subsistence production and because culturally they "have the responsibility of seeing that everyone in the household gets fed and cared for on a day-to-day basis" (81). As a result, it is women who act as 'conduits of genetic and cultural transmission' with women's knowledge of plant varieties and of food crop germplasm being of vital importance in preserving this knowledge before it is irretrievably lost.

Carolyn Sachs also points out that women's intimate involvement with plant production arises from the fact that "[I]n almost all locations, women cook and process food" (Sachs 1996, 72). Such involvement entails and fosters a detailed knowledge of what agricultural products are suited to particular ecosystems. The linkage between women's role as growers of food and as preparers, that is as cooks of this produce, is an extremely important one, even an obvious one, but one which is generally not sufficiently acknowledged and emphasised. Food crops have been developed over the millennia largely by women, who have an intimate hands-on knowledge of how these crops are best suited for particular culinary and health purposes. Therefore, the threat of the loss of genetic biodiversity is closely linked with the loss of indigenous and regional culinary knowledge and traditions, which have been developed by women with an intimate day-to-day knowledge of the local soil, climate and seasons. According to the *Global Biodiversity Strategy* put out by a group of United Nations organisations, including the United Nations Environment Program and the Food and Agricultural Organization:

> The loss of genetic, species, and ecosystem diversity both stems from and invites the loss of cultural diversity. Diverse cultures have bred and sustained numerous varieties of crop, livestock, and habitats. By the same token, the loss of certain crops, replacement of traditional crops with export crops, the extinction of species ...and the degradation or conversion of homelands are cultural as well as biological losses. (cited in Nazarea 2005, 8)

Included in these cultural losses is the loss of culinary knowledge or memory. The lack of recognition of this vital link is to do with the masculinist thinking that ignores or overlooks the vital role of women as holders of this genetic and cultural knowledge, and therefore either does not recognise, or trivialises women's role as producers of food and even more as preparers of food, failing to see the vital link between the two, a link which is at the heart of sustainability. This ignoring or disregard for women's expertise and knowledge is typical of androcentric thinking. Carolyn Sachs states that women's knowledge and expertise have long been overlooked by plant geneticists and seed collectors, with the locally adapted genetic traits of plants that have been selected over many generations by women growers being disregarded by agricultural scientists in favour of monocultures bred specifically for the market. She describes, for instance, the selection by African women farmers of maize varieties because of their agricultural and nutritional compatibility with beans.

Such intimate and detailed knowledge of environmentally and nutritionally sustainable multicropping, however, is disregarded or dismissed by agricultural scientists as backward, and standing in the way of modern agricultural development which emphasises "uniformity for easy, mechanized harvesting and increased yields" (Sachs 1996, 85). Consequently, "scientists and agriculturalists often consider women's knowledge about seeds, plants, and production techniques irrelevant and backward. They argue that 'scientific knowledge' must replace women's ways of knowing" (91). However, neglect by development programmes of women's detailed agricultural knowledge and their central role in food production in, for instance, Africa contributes to food insecurity and under-nutrition in these societies (Sachs 1996).

Nazarea and many others express deep concern at the alarming rate of genetic loss due to the displacement of highly biodiverse indigenous agriculture by modern industrial capitalist-driven monocultures. This substitution by genetically uniform cultivars has resulted in: "The rapid rate of destruction of crop variability [which] is in sad contrast to the rate of its creation – about 100 years compared to 10,000 years" (Nazarea 2005, 9). Since the adoption of Green Revolution agriculture in the Philippines, many farmers who are now in their fifties and sixties "are no longer knowledgeable about traditional methods of rice cultivation", and that as the older generation of farmers in their seventies and eighties are replaced by a younger generation, "much of this valuable cultural information generated and refined by long periods of coevolution will be irretrievably lost" (9–10).[2]

Another extremely serious threat to both biodiversity and cultural diversity comes from free trade agreements because these agreements favour large transnational corporations and agribusiness over the interests of small-scale local farmers and producers. In addition to the royalties charged by agribusiness who hold the patents on genetically modified (GM) crops, is the risk of litigation of farmers by biotech companies for the unintentional use of GM seeds caused by cross-contamination. This contamination, for example, poses a huge threat to the extremely rich genetic diversity of Mexican maize developed over thousands of years, and in turn to the entire way of life of the indigenous people, whose culture has been dependent upon the life-cycle of maize for millennia. Commenting on the central importance of corn to their entire way of life and their identity, and the danger to this posed by GM crops, a Huichol Indian farmer describes how: "Corn is the force, the life and the strength of the Huichol. If there were a change, if someone from outside patented our corn, it would end our life and existence" (cited in Carlsen 2004). Indeed, maize has come to epitomise some of the worst excesses of modern industrialised agriculture:

> Corn [has] adapted brilliantly to the new industrial regime, consuming prodigious quantities of fossil fuel...More than half of all the synthetic nitrogen made today is applied to corn, whose hybrid strains can make better use of it than any other plant. Growing corn...has in no small measure become a process of converting fossil fuels into food. (Pollan 2006, 45)

The Threat of Climate Change to Biodiversity and Cultural Diversity

Climate change or, more appropriately, climate chaos is probably the single greatest threat to biodiversity and hence to food security. Indeed, without biodiversity and its associated cultural diversity, human civilisation and its very survival is threatened. It is estimated that, because of its total dependence on fossil fuels, the industrial food system contributes between 44 and 57% of global greenhouse gas emissions; this includes the use of fossil fuel-based fertiliser, land-clearing for agriculture, transport, processing and packaging and food waste (GRAIN 2016). In fact, far from being the solution to world hunger, "our greenhouse–gas spewing industrial food system has entered a dangerous feedback loop. The way we produce and consume food is undermining our ability to produce food at all" (Holt-Giménez 2019, 3).

Vulnerability to crop failure as a result of climate change makes the preservation of genetic biodiversity of even greater urgency, as the vast variety of seeds and landraces that have evolved and been developed by farmers over millennia have been bred for their adaptability to particular climates and geographic environments. Throughout most of the countries of the Global South—in Africa, Asia and Latin America—women subsistence farmers produce between 60–80% of the food for local consumption (GRAIN 2016, 85). Their intimate knowledge of seed varieties that are adapted to their local climate and soil conditions means that they are the principal custodians of genetic knowledge and biodiversity (Nazarea 2005). However, this vast, invaluable and irreplaceable reservoir of genetic diversity is now under threat from modern industrialised agriculture, as well as climate change, of which industrialised food systems are a major contributor (GRAIN 2016). As a result of modern agriculture, especially since the advent of the Green Revolution in the 1960s, which promoted the use of hybrids at the expense of highly diverse indigenous varieties, the genetic base of agricultural crops in many countries has been drastically eroded. In the Philippines, for instance, where once thousands of different varieties of rice were cultivated, now just two varieties account for 98% of production, and in Mexico, 80% of maize varieties have disappeared since 1930 (Barker 2002, 314). This depressing scenario is being repeated throughout the world; according to a study by the FAO (Food and Agricultural Organization), 75% of the world's agricultural diversity was lost during the twentieth century (Lang and Heasman 2004, 222). Although 5,000–70,000 plant species have been documented as food for humans, only three—rice, wheat and maize—now provide half of the plant-based calories consumed by the world's population (Tutwiler et al. 2017, 3). Furthermore, despite the Green Revolution being touted as the answer to hunger in the developing world, during the course of the Green Revolution "the ranks of the hungry swelled by 11%" (Patel and Moore 2018, 151).

A study by Rural Advancement Foundation International found that only 3% of vegetable varieties that were available in 1900 had survived by the 1980s (Kimbrell 2002, 79), with a similarly drastic decline in the genetic diversity in breeds of farm animals. With meat, milk and egg production throughout the world becoming increasingly dominated by industrialised systems using only a small number of commercial breeds, "almost every month, another livestock breed becomes extinct" (Farming Matters March 2010, 44). This impoverishment of the genetic base is now being even further exacerbated by agricultural biotechnology, which has been developed in the decades since the 1980s. Pointing out that more than 90% of the world's commercially produced turkeys are currently derived from three breeding flocks (Lang and Heasman 2004, 223), British food policy academic Tim Lang warns of the dangers to human health through the risk of diseases resulting from the loss of genetic diversity, such as the dangers of a new virulent strain of avian flu developing in modern commercial breeding flocks that have no disease resistance due to their extremely narrow genetic base. Indeed, as evolutionary biologist Rob Wallace graphically describes in his important book, *Big Farms Make Big Flu*, the combination of genetic uniformity in industrial farm animal production—that is factory farms—with massively overcrowded and unhygienic conditions creates the perfect conditions for deadly pandemics. He points out that: "Anyone who aims to understand why viruses are becoming more dangerous must investigate the industrial model of agriculture and more specifically livestock production. At present, few governments, and few scientists, are prepared to do so" (GRAIN 2020). In fact, far from addressing the dangers, cruelty and horrors of factory farming, governments and regulators point the finger for pandemics such as avian flu and SARS at free-range animals raised on small farms, with the result that they are targeted for closure. This is very much in the interests of agribusiness as they want to eliminate any alternative forms of agriculture. Pandemics, such as avian flu, particularly affect small-scale farmers, who lose their livestock and hence their livelihoods. Consequently, small-scale farming and industrial animal production cannot coexist. However, globally it is big business that is promoted, protected and supported by most governments, at the expense of small-scale farmers producing food in a healthy and environmentally sustainable way.

The vital importance of biodiversity for the survival of future generations is emphasised by Virginia Nazarea: "In case of drastic and unforeseen changes in the environment, the presence of genetic variability on which natural selection can operate assures that there will be a perpetuation of species and even of life itself" (Nazarea 2005, 11). The preservation of biological diversity is fundamentally dependent upon the preservation of cultural diversity, "humanity's collective knowledge of biodiversity and its use and management resting in cultural diversity" (74). Furthermore, gene banks are of little use if the cultural knowledge that accompanies them and that has been acquired and transmitted over countless generations is irretrievably lost. Biodiversity and cultural diversity are interdependent, and therefore the preservation of cultural memory is as important as the preservation of genetic diversity. Nazarea emphasises the need for banking of both genes and cultural memory, that is, for: "the genetic, on the one hand, which embodies millions of years of biological evolution, and the cultural, which, on the other hand, embodies a long and rough – but often successful – history of human adaptation" (6). The negative impact of climate change is especially devastating for rural women in the Global South. Not only do they have to adapt to a changing climate, making food production increasingly difficult, but they often have to shoulder an even greater share of farm labour as men migrate to cities or other countries in search of work. Furthermore, in parts of Africa, food shortages and hunger caused by climate change are forcing young girls, some as young as thirteen to get married as their families can no longer afford to feed all their children (UN Women 2017a).

Global 'Supermarketisation'

Tragically, the mistakes that the industrialised West has made with regard to many aspects of food production—including the dominance of supermarket chains and the resulting displacement of small retailers and shop owners—are now occurring throughout much of the developing world, especially in countries such as China and India with their boom economies and rapidly growing population of middle-class consumers. For instance, in India, the livelihood of potentially millions of small shop-holders is threatened by the development of Western-style supermarket chains designed to "provide a welcome relief to India's young, aspirational middle-class seeking refuge from loud, crowded, dust-clogged markets" (Pepper 2006). In addition, the development of a supermarket culture inevitably leads to all the problems that are associated with supermarkets in the industrialised West, including environmental concerns, such as unnecessary packaging and an increase in 'food miles' contributing to greenhouse gases, as well as increased corporate control of the food supply.

The international organisation GRAIN, describing the 'supermarketisation' of the Asian food market, reports that trade liberalisation has promoted the rapid expansion of supermarkets in many Asian countries such as India, and China where supermarkets are growing by 30–40% per year, and Indonesia, where the survival of its 12.5 million stallholders operating in 13,450 registered fresh food markets, is threatened by corporate retail outlets increasing by 2000 per annum (GRAIN 2016, 196). This rapid supermarket expansion is destroying the livelihoods of millions of people in the local distribution chain including small traders, many of whom are women, who "sit at the front end of the local food systems that ensure the procurement and distribution of food grown on millions of small farms across the region" (197). Small farmers and traders are also particularly damaged by the expansion of supermarkets, because supermarkets require adherence to standards, developed in Europe and promoted by the FAO and Codex Alimentarius, and that demand precise standardised norms for size and appearance of fruit and vegetables "that are impossible for small farmers to comply with" (GRAIN 2016, 198). Vandana Shiva, speaking of the environmentally and socially destructive impact of agribusiness and the increasing corporatisation of the food supply in India, comments that "all the positive aspects of food – freshness, local supply, low cost, low environmental impact, and high nutrition

– are destroyed and replaced by negative aspects – staleness, long-distance supply, higher cost, high environmental impact, and low nutrition due to over-processing" (Shiva 2000, 88).

Traditional local healthy oils, such as mustard, coconut and sunflower, are being increasingly replaced by cheap industrially-produced soybean, palm oil and canola oil, which in 2012–2013 made up 76.6% of world's total oil production (GRAIN 2016, 201). This development is also facilitated by international free trade agreements, and tariff reductions imposed by the IMF and World Bank. As a result of such tariff reductions imposed on India in 2008, the income of Kerala coconut farmers was slashed by two-thirds, causing many to commit suicide (202). Furthermore, the global expansion of soy and palm oil monoculture production is causing disastrous environmental and social problems; these include land grabbing, the destruction of tropical rainforests to grow palm oil, adding to climate change, the displacement of peasants and indigenous people, and the poisoning of local populations caused by the aerial spraying of genetically modified soy crops with glyphosate and, as is well known, the threatened extinction of orangutans in Indonesia.

Stressing the importance of protecting both biodiversity and cultural diversity from 'mindless homogenization', Virginia Nazarea asserts that the multicropping systems of farmers of the Global South promote a wide variety of crops and food plants that support dietary diversity, as well as protecting against crop disease and providing a vital source of income with low expenditure due to the use of low-cost inputs and low-level technology. Diversity of diet is essential to human health. Australian nutritionist Mark Wahlquist points out that the relationship between biodiversity and dietary diversity is not sufficiently recognised and therefore not protected as it should be. He emphasises that the "evidence for food diversity as a major factor in health has grown substantially over the last few years", and that food diversity, which "provides the basis for food culture and cuisine" needs to be maintained and promoted at both the regional and intercommunity level (Wahlquist 2003, 304). There is a general positive relationship between food diversity and food security, and it is a mistake to regard food security solely in terms of available nutrients. In contrast, a "*food-based approach* to food security recognizes the need for a sustainable, affordable and diverse food supply, in *cultural context*" (305; original italics). And as outlined above, it is women who are pivotal to this, with "...diversity (being) best derived from biodiversity and women's links to it" (Mies and Shiva 1993, 164). Wahlquist also points out the importance of recognising women's central role with regard to food diversity and security, especially as women have historically been the inventors and developers of the technology related to basic human needs, including the technology to do with food-gathering, processing, storing and cooking (Wahlquist 2003, 307).

There is a common belief that food variety and choice in developed industrial societies is greater than at any time in history, and that the industrial wage earner has been emancipated from "a rather monotonous and regional diet" (Teuteberg 1975, 61). This is, however, largely a misconception; as pointed out earlier in this chapter, over the past century there has been an alarming decrease in agricultural crop varieties, with only three per cent of crop varieties having survived the last eighty years. The modern industrial American diet is based largely on two crops, corn and soy: corn is present in most foods, from the sweetening agent in soft drinks to the grain fed to industrially raised chickens, pigs and cattle, whilst soy is also fed to livestock, as well as being present in two-thirds of processed foods (Pollan 2006, 35). Indeed, the apparent variety available in modern developed nations is illusory.

> Even a superficial glance down supermarket shelves would seem to show that the American consumer has an abundance of choice when it comes to food choices. In reality, however, this is an illusion of choice rather than a real opportunity to experience food diversity. We certainly have a wide variety of brand names from which to choose. Unfortunately, the variety of corporate brand logos does not reflect biological variety. (Kimbrell 2002, 71)

The Homogenisation of Food Cultures

In his analysis of the American fast-food industry, Eric Schlosser describes how processed foods and the fast-food industry depend upon the use of artificial flavours to make its denatured food palatable. The fast-food industry, in recognising the connection between aroma and memory, and how a 'smell can evoke a long-forgotten moment", especially from people's childhood, goes to great lengths to exploit this:

> The flavours of childhood foods seem to leave an indelible mark, and adults often return to them, without always knowing why. These 'comfort foods' become a source of pleasure and reassurance, a fact that fast-food chains work hard to promote. Childhood memories of Happy Meals can translate

> into frequent adult visits to McDonald's, like those of the chain's 'heavy users', the customers who eat there four or five times a week. (Schlosser 2001, 123)

This exploitation by the food industry of an intrinsic human need to be reconnected through the senses to the past via the use of a vast battery of artificial tastes and smells creates in effect an ersatz world of the senses—in other words, an inauthentic and disconnected experience. The unconscious dissatisfaction resulting from this, accounts, I would suggest, to a large extent for the hyper-consumption of modern society referred to by Nadia Seremetakis:

> Each episode of consumption is relatively absolute and quickly totalising because it never lingers long enough in the senses as social memory...To be stitched into a historical fabric with the others it has displaced...In the high turn-over of commodity experiences, each object, each material experience, is the absence of the other...(which) allows (for) hyper-consumption. (Seremetakis 1994, 10)

Nutritionist Marion Nestle points out that the very abundance of US food (enough to feed everyone in the country twice over, after exports are included) means that the challenge for food companies competing against each other is to persuade people to eat their particular product and, furthermore, to *eat more* of it. Despite the long-standing health advice for people to eat healthily and in particular to consume less fattening foods, food companies are in the business of maximising profits for their shareholders, and therefore they "work hard to oppose and undermine 'eat less' messages", seeking out "new audiences amongst children [and]...minority groups, or internationally" (Nestle 2002, 21). In order to increase profits, food companies have introduced a new strategy, "increasing the size of food portions" (21). As a result, the 'super-sizing' of fast-food portions has resulted in, for instance, a threefold increase in the number of calories in a serving of McDonald's French fries (Critser 2003, 28). The rise of high-calorie, high-fat fast-food in the United States started in the 1970s when the Secretary for Agriculture, Earl Butz, promoted the use of high-fructose corn syrup as a major component in the cheap mass-production of processed foods, and an increase in meat production supported by low-cost subsidised soy meal and corn as feed for factory-farmed animals. As a result of this strategy, at "the supermarket, calorie-dense convenience foods (became) more and more affordable" (18). Consequently, "Americans are the fattest people on the face of the Earth" (4), with more than 60% being overweight and one in five being obese. Although obesity is increasingly affecting middle and upper income groups, this is a health crisis which particularly affects the poor, especially low-income Blacks, Hispanics and the poor white of areas such as the Appalachians: "Poverty. Class. Income. Over and over, these emerge as the key determinants of obesity and weight-related disease" (116). Fast-food companies often target as their market poor immigrant populations such as Latinos, who are also metabolically at particular risk of obesity. This arises from a long colonial history of cultural and political oppression, resulting in many generations of children born to mothers who are nutritionally deprived. Once such populations are exposed to high-fat industrialised foods they rapidly become obese because their metabolism has evolved in a 'culture of scarcity' and hence programmed to store fat. As one anthropologist studying this phenomenon comments: "In a strange way, you are seeing the original trauma of (colonial) conquest played out, metabolically, in the streets of Los Angeles" (cited, 130).

One of the consequences of the overabundance of the American food supply is "pressure to add value to foods through processing" (Nestle 2002, 21). Food companies' profits come from the processing of raw materials, with the price being paid at the supermarket increasing in proportion to the amount of processing; yet the primary producers only receive a small fraction of the total price, with "the proportion represented by the farm value declining in proportion to the extent of processing" (17). American food policy expert, Wenonah Hauter, observes that 90% of the food budgets of Americans is spent on processed foods, and half of meals are eaten away from home, "increasingly at fast-food restaurants" (Hauter 2012, 66). British journalist, Joanna Blythman, refers to the year-round availability of supermarket produce, regardless of season, as 'permanent global summertime', which, she says, leads shoppers in the United Kingdom "to expect virtually every horticultural product on the market every day" (Blythman 2004, 74). Much of this demand is satisfied by the importation of foods from developing countries where traditional subsistence agriculture, often done by women growing food for local consumption, has been displaced by the export of cash crops grown in order to pay off the crippling debt incurred by loans from the World Bank and the International Monetary Fund. In addition, such a food system is not only socially and economically destructive, but is also environmentally unsustainable, with an absurd amount of transport involved; for instance, according to the British lobbying group, Sustain, the average British Sunday lunch might include (factory-farmed) chicken from Thailand, beans from Zambia and carrots from Spain, entailing a transport distance of more than 26,000 miles.

The Nutrition Transition, Nutritionism and Biofortification

Since the Second World War, and especially since the 1970s and 1980s, the rapidly increasing industrialisation of the food system and modern diet, coupled with an increase in sedentary lifestyles due to technology, has "clashed with out biology to create major shifts in body composition" (Popkin, et al. 2012). The negative consequences of this transition—obesity and related health problems, including cardiovascular disease, diabetes and cancers—initially affected industrialised Western countries, particularly the United States, and to a lesser extent Europe. However, in the last two to three decades this shift in diet, referred to as the 'nutrition transition' has also become a major problem in low- to middle-income countries, including China, India, Brazil, Mexico and other Latin American countries, as well as the Middle East (Popkin et al. 2012; GRAIN 2016).

The 'nutrition transition' refers to a modern industrialised food system and diet and "is broadly defined by high intake of refined carbohydrates, added sugars, fats and animal-source foods" with traditional diets "rich in legumes, other vegetables, and coarse grains disappearing in all regions and countries" (Popkin et al. 2012). One unexpected example of this negative development is that of Lebanon, whose traditional grain and vegetable-based diets is one of the healthiest, most varied and delicious in the world. However, this traditional diet based on fresh, local ingredients including vegetables and wild plants, grains, legumes and fruit, with a moderate amount of animal-based foods, is being "eroded in favour of a Westernized low-fiber, high fat, high sugar diet…accompanied by increases in the rates of various chronic diseases" (Batal and Hunter 2007). Unfortunately, as countries become more economically developed, the desire, especially amongst young urban people, to eat Westernised diets increases, with traditional diets often being stigmatised (Gold and McBurney 2012). Significantly, contrary to the popular image of starving

Africans, an in-depth study found that the diets of a number of African countries, including Sierra Leone, Cameroon, and Guyana, as well as Sub-Saharan Mali and Chad were amongst the healthiest in the world (Wilson 2019, 28). This is because, although food may not be overabundant, as in the developed world, the traditional healthy diet which still prevails is based on unrefined grains, such as millet, sorghum and maize, with plenty of greens, providing a diet rich in micronutrients (30). Wahlquist remarks that, with regard to food variety, and therefore health status, in fact the traditional cuisines of many so-called 'developing' countries have an advantage over those of much of the 'developed' world. Marion Nestle concurs, stating that the traditional plant-based cuisines of many developing countries are amongst the healthiest in the world, as well as being some of the world's finest cuisines. As long as calorie intake is adequate, people raised on such diets, once they have survived infancy, tend to enjoy "an adulthood relatively free of chronic disease until late in life" (Nestle 2002, 16).

In her important book analysing the connection between obesity and social inequality, Julie Guthman points out that the rapid increase in obesity and diet-related diseases started around the early 1980s with the introduction of neoliberal capitalism which deregulated environmental and social safeguards. She asserts that "it is the failure of environmental and food health and safety regulations that is most directly traceable back to neoliberal policy and forward to obesity" (Guthman 2011, 178). Furthermore, the neoliberal economy depends upon a cheap, underpaid labour force, many of whom work in agribusiness and the food industries, and who in turn have no option but to buy the cheap, poor quality products of these industries, thereby achieving a neat outcome for neoliberal capitalism: an underpaid workforce which provides a market for junk food. One of the factors implicated in industrially produced food resulting in obesity, is the presence of endocrine disrupting chemicals (EDC), referred to as 'obesogens' including in pesticides and herbicides, the use of soy to fatten cattle, and the materials, particularly plastics, used in processing and packaging. Consequently: "It appears that obesogens are present all along the food supply chains from farm production to transportation and storage to food processing" (Guthman 2011, 109). Therefore, the high incidence of obesity in economically and socially disadvantaged populations is because "low-income people eat more foods produced with the most obesogenic-intensive industrial farming and food-processing methods *because they are cheaper*" (112, italics added).

The rise of supermarket culture which promotes nutritionally deficient, industrially processed foods, and the resulting decline in local markets selling fresh food, is causing a huge change in the dietary habits of many so-called developing countries. This in turn is causing a massive increase in many diet-related diseases, such as cancers, diabetes and heart disease, as well as an explosion in obesity. A GRAIN report (2016) found that the increase in obesity is especially marked in countries with the greatest expansion of supermarkets and increase in urbanisation, resulting in a dramatic shift away from the consumption of grains and complex carbohydrates towards the consumption of meats and fats such as palm oil in processed and packaged supermarket products. For instance, in China where more than half of the 1.3 billion population now live in urban areas, the number of obese people in China has risen from 18 million in 2005 to 100 million in 2011, with similar trends occurring in India, Pakistan and Indonesia (203).

The relatively cheap price of processed 'junk' food compared with fresh fruit and vegetables means that it is especially the poor who are increasingly dependent on an unhealthy, nutrient-poor diet. Nowhere is this more true that in Mexico which "provides a stark and horrific picture of the consequences for people ....(where) poverty, hunger, obesity and disease go hand in hand" (GRAIN 2015). Free trade agreements have had a devastating impact on Mexican food security, where it is estimated almost 70% of the population is food insecure (GRAIN 2015). The North American Free Trade Agreement (NAFTA) in particular, which was signed in 1994, triggered a massive expansion in Mexico by US food processing companies, such as Coca-Cola and Pepsico. With the rapid spread of supermarkets and the takeover of traditional *tiendas* (local small food shops) by foreign companies such as Coca-Cola and Walmart, Mexico became flooded with cheap junk food and soft drinks.[3] The intention of such companies was "the absolute domination of the sales point", giving people, especially the poor, no option, with the only available goods being mostly processed foods (2015). Consequently: "Mexico has one of the highest rates of obesity, diabetes and high blood pressure in the world. It is the world leader in consumption per head of cola drinks and one of the highest consuming countries of 'junk food'. The consumption of maize products in Mexico has also started to decline for the first time in history" (cited, GRAIN 2015, 9). To add insult to injury, in an attempt to address the chronic food insecurity and health problems caused by an increasing diet of industrially processed food, the Mexican government signed agreements with Nestlé and PepsiCo, involving them in a 'National Campaign Against Hunger' (2015). Included in this was a project aimed at 'women entrepreneurs' called *Mi Dulce Negocio* (My Sweet Business) with the eventual aim of training a 'small army' of 15,000 women across Mexico to make sweet, but 'nutritious' desserts for children containing Nestlé products (2015).

The denaturing of food by corporate industrialised processing has resulted not only in an obesity 'epidemic', but also widespread malnutrition, the two problems often occurring in the same person. This is a serious problem that particularly affects economically and socially marginalised populations. Instead of addressing the root cause of the problem, capitalism's response is to view this as a marketing opportunity to increase profits, by bio-fortifying foods with the vitamins and minerals that have been lost by industrial processing. This reductionist approach to food is termed 'nutritionism' by Australian food academic, Gygory Scrinis, who describes it as:

> The nutritional engineering of foods (as) essentially a narrowly framed technological solution – or nutritional techno-fix – to the perceived limitations of particular foods or dietary patterns. If the problem is that highly processed foods have been stripped of beneficial nutrients, then engineering them back into foods is a way of convincing consumers to continue eating these products. (Scrinis 2013, 200–201)

In short, nutritionism and biofortification are technical fixes to a political problem, an opportunity for agribusiness to profit from dietary deficiency, much of it caused by the industrialised food system itself. Micronutrient deficiency, which affects 2 billion both in the Global North and South (Holt-Giménez 2017, 191) is often referred to as 'hidden hunger', the effects of which "include child and maternal death, physical disabilities, weakened immune systems, and compromised intellects" (cited, 191). Instead of acknowledging that the real solutions to hunger and nutrient deficiency are structural adjustment measures like land reform, the promotion of agroecological methods of agriculture, market reforms, and living wages, biofortification "allows governments and industry to depoliticize the causes of world nutrient deficiency by casting them as technical problems to be solved by technical solutions" (195).

One of the most controversial, if not notorious, biofortification projects has been that of Golden Rice, which was developed in 1990 and launched by the agrichemical company, Syngenta in 2000 (GRAIN 2018). Golden Rice was developed by using genetically engineering technology to produce rice containing beta-carotene, whereby bacteria and the genes from daffodil and maize were inserted into rice, giving it its golden colour. The professed purpose was to prevent vitamin A deficiency which causes blindness in "an estimated hundred thousand children annually in countries such as the Philippines, Bangladesh, Indonesia and India" (2018, 3). Despite twenty years of promotion, research and development, Golden Rice has failed to meet with acceptance from small-scale Asian farmers resisting GM technology, or to be an adequate source of vitamin A. In fact, according to the Philippines-based International Rice Research Institute (IRRI) "Golden Rice contains less than 10% of an equivalent amount of beta-carotene in carrots (and) with one carrot having the same amount of vitamin A found in nearly 4 kg of cooked GM golden rice", and one sweet potato having almost fifty times of beta-carotene as the equivalent amount of golden rice (12). Despite the discrediting of the beneficial qualities of Golden Rice, it continues to be promoted by the biotech industry as the 'poster child' for the promotion of GM foods to a sceptical public as the solution to world hunger, and even cynically accusing opponents of not caring about the world's hungry. Significantly, one of the main reasons for this vitamin A deficiency is the introduction of chemicals in agriculture, particularly herbicides, such as glyphosate that kill the nutritious 'weeds' and wild plants that are an important traditional source of nutrients, including vitamin A, in many non-industrialised countries. Instead of addressing the real cause, the reductionist biotech solution is Golden Rice. Consequently, claims that the solution to hunger can be solved by techno-fixes, such as GM foods and biofortification of foods, masks how the globalised industrialised food system has "destroyed traditional sources of nutrients and impoverished people's diets" (Holt-Giménez 2017, 194). The real solution to malnutrition lies in a nutritious and varied diet, because, unlike supplements, a variety of different foods consumed together combine and interact with many beneficial substances that protect against disease (Gina Kennedy et al. 2017, 24–27).

American women's studies academic, Aya Hirata Kimura makes the very important point that "when hunger is perceived as a technical problem to be solved by experts, the poor – particularly impoverished women – are marginalized" (Hirata Kimura 2013, 18). Moreover, instead of addressing the causes of poverty and food insecurity, biofortification programmes promoted by industry and organisations, such as the World

Bank, USAID and UN organisations including the Food and Agriculture Organisation (FAO) and World Health Organisation (WHO) are "inevitably accompanied by the scrutiny of mothering practices. Concern for the welfare of children ….often results in blaming mothers as ignorant, but it rarely translates into commitment to mothers' welfare and improvement in their overall living conditions" (27). In short, by promoting biofortification as the solution to malnourishment, neoliberal policies deny women's ability to nourish their children, and opens the market to corporate 'solutions'. It is also, by implication, a dismissal or lack of recognition of sustainable food systems and women's traditional food and nutritional knowledge. Additionally, it treats women as victims of their own biology. As Hirata Kimura so incisively explains:

> Women's identification as vulnerable victims of micronutrient malnutrition has relegated them to the position of passive *recipients* of food aid and nutritional expertise. Such a depiction of women as passive and needy victims risks reinforcing a gender stereotype that portrays women as inherently weak and powerless. The governments and experts become the active, benevolent 'doers' of things, as they further reach into the lives and bodies of women…..The universalizing identification of biological victimhood hinders a more feminist strategy for increasing women's participation in food policymaking… (Hirata Kimura 2013, 60–61; original italics)

Far from viewing women as victims whose 'inability' to adequately nourish their families is a 'problem' that needs to be solved by corporate market-driven strategies, such as biofortification, the solution to hunger and nutrient deficiency in fact lies in recognising and respecting women's traditional knowledge of sustainable food production and preparation that still exists in much of the world, and which has resulted in innumerable healthy and delicious regional cuisines. In a world where 2 billion suffer from micronutrient deficiencies and 1.9 billion are overweight or obese (FAO), the current industrial food system has not only failed to provide adequately for the world's food needs, but has in fact been a major cause of hunger, malnutrition and obesity, as well as causing disastrous environmental damage, particularly loss of biodiversity and climate change. Therefore, the answer lies not in more of the same as promoted by agribusiness corporations and most governments. The real solution to hunger, as well as to environmental degradation lies in supporting agroecological food production practised by small-scale peasant farmers, many of whom in the Global South are women who produce most of the food for local consumption, as well as being the principal promoters and guardians of biodiversity.

Land Grabbing

There are many threats and challenges that small-scale farmers face in today's world. Apart from probably the most serious threat of climate change, these threats include land grabbing. The appropriation of small-scale farmers' lands for corporate/capitalist profit has a long history, going back to the first English Acts of Enclosure during the sixteenth and following centuries (as I described in Chapter 6), as well as centuries of displacement of indigenous peoples driven by European colonialism. However, it is only in recent decades that land appropriation, referred to as 'land grabbing' is occurring on a global scale. Indeed, the "new land rush looks like a final enclosure of the planet's wild places, a last roundup on the global commons" (Pearce 2012, intro.).

The 2008 global financial crisis and the food crisis caused by rising prices, coupled with global warming and rising world population created a 'perfect storm', leading to a rush on the part of many wealthy, but food-insecure countries to purchase or lease vast tracts of land for agriculture especially in countries of the Global South. Land grabbing is driven by food-insecure countries, such as China, South Korea and Japan, as well as the Middle East Gulf States, such as Qatar and the UAE (United Arab Emirates) whose governments are anxious to secure food supplies in order to avoid food shortages and resulting social unrest (Shepard and Mittal 2009). Countries from the Global North are also involved in land grabbing. For instance, the US Boston-based Hancock Agricultural Investment Group (HAIG) has purchased almost 44,000 hectares of agricultural land in Australia since 2000, mainly for almonds, macadamia nuts and cotton (Tracey 2020). Other motives behind land grabbing are growing crops for biofuels, and land speculation by wealthy individuals, such as George Soros, Richard Branson and Lord Rothchild (Pearce 2012, intro) as well as corporations, banks and private investment and public pension funds.

Although land grabbing is a global phenomenon, it is especially prevalent in the Global South. Governments and organisations driving land grabbing argue it is in the interests of poor countries to adopt industrial agriculture, claiming that that is the only way that the world can be fed, and that there are "still many areas of the world – including large swathes of Africa – that have good land that could be used far more productively" if given over to industrial agriculture (Pearce 2012, 294). The production model aggressively promoted by land grabbing is that of industrial monoculture agriculture largely for export "to the detriment of family farmers, the environment, and the food and nutritional security of the world's poor" (Wise 2019, 9). As Fred Pearce asserts: "although simple measures of tons of grain per acre may suggest big is best", the purpose of big farms is "to generate returns for capital – to please investors, rather than to feed families" (2012, 295). In contrast:

> [S]mall farmers bring many other things to the kitchen table. Official statistics often ignore the fact that they use every corner of their plots, planting kitchen gardens where mechanized farms have vehicle yards. They gather fruits from hedgerows. They have chickens running in the yard. They feed animals on farm waste and apply manure to fields. They raise fish in their flooded paddies. (295)

This is precisely the approach to farming that millions of small-scale women farmers practise throughout the world, as I describe in detail in the next chapter. As Pearce points out, what is needed is investment in small-scale farming, rather than industrial agriculture which "hollow out communities ...and turn small holders into laborers for corporate farms" (296). He gives the example of Vietnam, which, by investing in smallholder farming, has gone from being an importer to an exporter of food.

One of the most serious consequences of land grabbing is the clearing/destruction of forested and wilderness areas, such as that in the Amazon, for both beef cattle farming for export, and where soy is grown on an industrial scale for export as feed for animals in intensive production systems. Africa, in particular is seen as the 'new frontier', especially West Africa, such as the Guinea Savannah Zone that the World Bank describes as "one of the largest underused agricultural land reserves in the world" (cited, Wallace 2016, 327). A serious consequence of wilderness destruction is the increasing threat of pandemics. In his important book on the links between industrial agriculture and pandemics, such as SARS and avian flu, evolutionary epidemiologist, Rob Wallace analyses how the destruction of vast swathes of wilderness in West Africa for industrial palm oil production was the probable cause of Ebola outbreaks in recent years. The destruction of the forest for palm oil plantations has severely disrupted local ecosystems; as a consequence of their habitat being destroyed, certain species of bats that are 'key reservoirs for Ebola' seek refuge in palm plantations where workers are exposed to the Ebola virus (Wallace 2016, 329). According to Wallace, "nearly every Ebola outbreak to date appears to be connected to capital-driven shifts in land use, including logging, mining, and agriculture" (330). He also makes the interesting and important point that the origins of such pandemic outbreaks might actually be considered to be the seats of capital, such as New York, London and Hong Kong, as neoliberal capitalism is the driver of industrial agriculture and ecological destruction that create the conditions for pandemics such as Ebola, SARS and Avian flu.

Land grabbing has a devastating impact on affected rural communities. Women in particular are often the most affected by this in 'developing' countries, as these acquisitions are often accompanied by violence (Ross 2019). For instance, in West and Central Africa where governments have given over four million hectares of land for foreign-owned industrial palm oil and rubber plantations, local women and girls are exposed on a daily basis to sexual violence, assault and rape from company workers and guards. Women who have been dispossessed of their own land for growing food, and hence their livelihoods, are forced into low-paid work in the plantations where they are routinely sexually abused; so-called contractor heads regularly demand sex before giving jobs to women, or before paying them for work already done. As one Liberian woman says: "If you are unlucky you only get paid if you let the guy 'do his thing'. It happens all the time" (GRAIN 2019, 2).

The loss of land and the accompanying violence against women also has serious consequences regarding food security, as women produce almost 90% of food in the countries of Sub-Saharan Africa (2019, 4). Deprived of their land for food production, women are forced to "try and grow some cassava and beans along the pesticide-infested fringes of industrial plantations" (4). Furthermore, they can no longer grow the local traditional non-industrial palm oil which is an important ingredient in the local diet, as well as use as a medicine. When they do manage to grow or gather a few palm nuts, plantation guards confiscate them, and even invade the women's kitchens in search of them. Unsurprisingly, no action is taken against the abusers by the company or the local police. In one case, where a ruling was made in favour of the victim, the Belgian-French-owned company, SOCAPALM, appealed against it. In addition, women have been deprived of an important source of independent income from the processing of traditional palm oil. Despite women's dependence on access to land for their subsistence livelihoods, "traditional customs often still exclude them from decision-making over the use of the land they depend on to feed their families" (7). Nevertheless, or perhaps because of this, women are at the forefront in the fight for the defence of their land. A particularly negative aspect of land grabbing is that it is usually the most fertile and well-watered land that is seized, resulting in either indigenous farmers being pushed on to less productive land or being totally dispossessed of any land.

The so-called Livestock Revolution, which promotes industrialised animal production, has caused the dispossession of many small-scale family farmers in both the Global South and the Global North. As a result of having the most corporatised livestock industry in Asia, principally poultry and pigs, by the mid-1990s farmers in Thailand had become contract farmers for the booming export market, essentially becoming indentured factory workers for agribusiness. Consequently: "For the majority of Thai farmers, the Livestock Revolution has meant soaring indebtedness, loss of independence, and the continued migration of their daughters to Bangkok's sweatshops and brothels" (Davis 2005, 99).

Food Security, Food Sovereignty and Agroecology

The UN FAO states that: "Food security exists when all people, at all times, have physical, social and economic access to sufficient, safe and nutritious food which meets their dietary needs and food preferences for an active and healthy life" (FAO 2008, 3). While agencies, such as the FAO, are primarily concerned with food security, as defined above, in 1996 the concept of food sovereignty was put forward by NGO, especially by La Via Campesina peasant's movement who argued that "there could be no food security without *food sovereignty*" (Patel 2011, 188; italics added). Whereas the original definition of food security could mean the United States, for instance, dumping its surplus grain on Third World countries (as they attempted with GM maize in Zambia), food sovereignty promotes and protects the rights of producers to have access to their *own culturally appropriate foods*. According to La Via Campesina, food sovereignty also means ecological and sustainable methods of food production, ensuring that the "rights to use and manage their lands, territories, waters, seeds, livestock and biodiversity are in the hands of those who produce food" (190).

Of central importance to the concept of food sovereignty is *seed sovereignty* because whoever "controls the seed gains a substantial measure of control over the ...entire food system" (Kloppenburg 2011, 152). Until the 1930s, throughout the world, farmers had nearly complete seed sovereignty, meaning they decided what seeds to plant, which ones to save and with whom they shared them, either for food or for planting (153). This 'open system' was governed by reciprocity rather than by the market; indeed, "these customary arrangements usually functioned to stimulate and facilitate – rather than restrict – the wide dissemination of seed" (153). However, since the 1930s farmer's seed sovereignty has been undermined by the development of the 'life sciences industry', with control of seeds increasingly being taken over by private capital for profit, such as through the hybridisation of crops, as well as legislation that gave exclusive control to industry over crop varieties that could not by hybridised (154). The increasing globalised corporate control of seeds has developed apace over the last three decades with the development of genetically modified organisms (GMOs). The increasing corporate takeover of seeds, including through the patenting of seeds, is an attack on farmers' rights to save seed which is fundamental to food security and food sovereignty.

Integral to food sovereignty is the practice of agroecology, which is "an integrated approach that simultaneously applies ecological and social concepts and principles to the design and management of food and agricultural systems". Furthermore, it aims "to optimize the interactions between plants, animals, humans and the environment while taking into consideration the social aspects that need to be addressed for a sustainable and fair food system" (FAO 2019, 1). In contrast to industrial monoculture agriculture, agroecology systems are biologically diverse, incorporating agroforestry systems, intercropping, crop rotation and integrated crop-livestock systems, and traditional fish polyculture, such as in 'fish-friendly' rice cultivation. Indeed, agroecology, with its emphasis on biodiversity, is now recognised by the FAO as the best way to provide the world's increasing population with nutritious diets, whilst also contributing huge environmental benefits, including soil health and encouraging pollinators which are crucial to agricultural production. Moreover, agroecological food production makes a major contribution to tackling climate change, as the use of natural fertiliser, such as manure and compost, encourages healthy plant growth, drawing down carbon from the atmosphere into the soil. This stands in stark contrast to the fossil fuel-based fertilisers of industrial agriculture that are a major source of greenhouse gases, and which result in depleted, hydrated soils that need more and more chemical fertilisers to be productive. Agroecology treats nutritious and culturally appropriate food as a human right, in contrast to globalised food markets promoted by organisations such as the World Trade Organisation (WTO) which treat food simply as a commodity. As Peter Rosset explains: "Food *is* different. It is not just any merchandise or commodity. Food means farming, and farming means rural livelihoods, traditions and cultures, and it means preserving, or destroying rural landscapes" (Rosset 2006, 9; original italics). According to the International Panel of Experts on Sustainable Food Systems (IPES-Food), "instead of the uniformity imposed by globalised markets, agroecology promotes diversity (from cultivated plot to plate, from the local to the global), which enables citizens to regain autonomy over the flows that link production to consumption at local or territorial levels" (Farming Matters 2018).

It is often argued/assumed that industrial agriculture is needed to feed the world. However, despite the fact that corporate agribusiness is the dominant mode of food production promoted by most governments and development agencies throughout the world, 70% of the world's food is still produced by over 2.5 billion small-scale farmers (Pimbert 2018). Agroecology is at least as productive and often more so, apart from the important fact that it also provides far more nutritious diets, as well as being environmentally sustainable and regenerative. Agroecological expert, Miguel Altieri, asserts: "Although conventional wisdom claims that small family farms are backward and unproductive, research shows that small farms are much more productive than large farms if *total* output is considered rather than yield from a single crop" (Altieri 2011, 122, italics added). Agroecology aims to replicate natural systems in a 'closed-loop', non-waste system with a:

> bottom-up construction of knowledge and grass-roots innovations that need to be *supported* – rather than led – by science and policy. Modern principles of agroecology have their roots in the collective knowledge, practices, and ecological rationale of indigenous and peasant agriculture(s) throughout the world. (Pimbert 2018, 53; italics added)

These practices and knowledge are based on centuries of collective observation and adaptation involving sophisticated, dynamic and complex management practices (2018). Of great importance is the recognition that much of this collective knowledge is that of women peasants and small-scale food producers. Indeed, because of their role as the primary food producers in much of the world, especially the Global South, women have been central to shaping the concept of food sovereignty (*LEISA* 2009).

The botanical knowledge of indigenous and peasant women farmers is of vital importance to the promotion of food sovereignty and agroecology because, as ethnobotanist, Patricia Howard asserts: "today and in the future, the way we view women and gender relations in relation to the plant world will greatly influence our ability to halt the erosion of plant biodiversity across the globe, particularly those plants that humans have found to be useful" (Howard 2003a, 2). Collectively, women are the principal holders of most of the world's plant knowledge. This is because "throughout known history, women's daily work has required more of this knowledge than men's daily work" (Howard 2006, 2). With the world facing unprecedented threats and challenges to food supply as a result of climate change, depletion of resources, including reduced soil fertility, a drastic reduction in biodiversity, and a projected population of more than nine million by 2050, a radical change to food systems needs to take place. Key to this is agroecology and the role of small-scale and peasant farmers, especially women farmers. Women are in fact central to agroecology; indeed, there can be no real agroecology without women because agroecology is about gender equality and social justice as well as about sustainable and regenerative food systems and environmental protection. Given women's historical and current role in sustainable food production, including seed saving and the promotion and preservation of biodiversity, it is essential that women are not only included but at the *forefront* of agroecological design and practice, with "the concept, vision and practice of agroecology, as promoted and defended by women (requiring) a recognition and appreciation for the practices and knowledge of women and their ancestors cultivated over centuries and passed from generation to generation" (CSM 2019, 4). Furthermore, women's knowledge must be valued by the agroecology movement and incorporated into agroecology education, research and development. Indeed, "agroecological practices and policies without the participation of women as central protagonists is not an option" (2019, 15).

A feminist ethic of care for the environment and communities underpins agroecological practices and is a direct rejection of the masculinist environmentally destructive and socially unjust ideology of industrial food systems. Nevertheless, the emphasis of small-scale family farming which is at the heart of food sovereignty and agroecology often "idealises family farming and the culture of rural and indigenous communities" (Montiel et al. 2020, 8) without addressing the oppressive and deeply unequal relations and power imbalance that exist between the sexes in a world based on male supremacy. To counter this, the women of La Via Campesina realised the need "to create their own assembly within the organisation to fight for their participation and to ensure that feminism was taken up as *everyone's* issue" (8, italics added). Based on this, feminist agroecology is about "generating a new solidarity economy where productive and reproductive work is made visible and shared, and where the patriarchal family structure is de-constructed and rebuilt based on *free and equal relations and shared responsibilities*" (CSM 2019, 7; italics added). Women bear the huge burden of the taken-for-granted, unpaid and undervalued workload of the reproductive domestic sphere, in addition to the responsibility of being the producers, promoters and the "custodians of healthy and culturally-based food practices that build on the productive wealth of local agriculture" (12). Therefore, a feminist agroecology demands that within the family men share care work which is "vital for both human life and the planet" (13) This involves child care and housework, such as cleaning, and not just the creatively satisfying activity of cooking that men often appropriate, as I have described in Chapter 4.

A feminist agroecology also demands that women have the same access that men have with regard to support services, including access to markets, credit, research and training, and, most of all, *the right to land*. It is a shocking fact that despite being the main producers of food in many countries, very few women own the title to the land they farm, and what land they do farm on is often not large enough and of inferior quality. Despite providing 70% of the world's food, small-scale peasant producers own only 20–30% of the world's farm land (GRAIN 2014); the rest is owned and controlled by agribusinesses producing commodity crops on mega-farms, mostly grains and oilseeds, much of it as animal feed for industrialised animal production, as well as for biofuels. It is often claimed that the estimated 1.6 billion peasant women of the world own only 2% of farmland (Korol 2016); however, this may be an underestimate, and it is sometimes disputed (Doss et al. 2020). Nevertheless, it is a fact that female peasants, despite their major role in food production, as well as doing much of the work of the family farm as a whole, have very little guaranteed access to land; this is due to lack of money or access to credit to purchase land, but also the fact that in many countries, especially in much of Africa and South Asia, patriarchal customary rights deny women the right to inherit land (Murali 2020).[4] Although women peasants generally own very little land, much of their food production is carried out via access to land owned by male members of the family, or on common land. The importance of women having secure rights to land cannot be overestimated. According to UN human rights reports, when women do have such rights, "myriad benefits tend to follow...(including) enhanced status, improved living conditions, better nutrition and food sovereignty, improved health and education outcomes, higher earning and individual savings, and better access to credit, as well as better protection from gender violence" (UN Women 2017b).

In addition to supporting women farmer's central role in producing nutritious and diverse food systems, agroecology has many other benefits for female producers: it helps to challenge familial and social patriarchal power structures, encouraging networking between women and women-only spaces "which are of great importance for achieving gender equity, building solidarity and autonomy, and strengthening women's creative and collective work towards self-determination" (CSM 2019, 6). It also creates better economic opportunities for women, being far more productive than chemically dependent industrial monocultures (CSM 2019; Mpofu 2018). Agroecology is the best means of contributing to the Sustainable Development Goals (SDGs) of "eradicating poverty, ending hunger, achieving food security, promoting nutrition and sustainable agriculture and ensuring that no one is left behind" (Mpofu 2018, 33).

Unfortunately, the very success of agroecology means that it also runs the risk of appropriation, both by men at the expense of women, and by agribusiness. A striking example of male takeover of agroecology from women is described in the case of the Basque Country, where, despite the fact that women were the ones who initiated agroecology, as it became more economically successful it attracted more and more men, with the result that women's participation in training fell from 80% to 20% (Vispo et al. 2020, 18). In Colombia, women peasant farmers, who had practised agroecology for decades, promoted it as "a tool for peace building, because as a social movement it helps strengthen rural organisations and improve peasants' living conditions" (19). This approach was adopted eventually at a state level; however, again women were sidelined with "women's participation and proposals often misused and misunderstood to promote ....the use of technological ...production for the exclusive export of 'exotic' foods" (20). The scaling up and out of agroecology is vitally important; however, as soon as it becomes economically successful it runs the risk of being appropriated and made to fit the capitalist patriarchal model (19), as has often happened with organic agriculture (Guthman 2014). This appropriation and misuse of women's knowledge and work underscores the "need to build new paradigms for agroecology in which women are visible and feminism is a priority" (Vispo et al. 2020, 20).

Biodiversity and cultural culinary diversity have evolved out of women's historic role, not only as cooks and preparers of food, but also traditionally as producers of food throughout the world: homegardens, which represent some of the richest cultivated biodiversity on the planet (Howard 2003b), and small-scale mixed farming, incorporating plants and animals, such a chickens, pigs, goats and milking cows. In the following chapter I examine the intimate relationship between cultural culinary diversity and biodiversity, and women's central role in developing and sustaining it. I examine women's detailed and intimate knowledge of plants and their culinary use, as well as women's important role in animal husbandry. Small-scale women farmers through their agroecological practices are the main custodians of seeds and of plant breeding. An acknowledgement and respect for this female cultural heritage, and support for small-scale farmers throughout the world, the majority of whom are women especially in the Global South, is the *real* solution to solving world hunger.

Notes

1. Is it worth noting that Taco Bell eventually failed in Mexico https://www.mashed.com/234254/why-taco-bell-flopped-in-mexico/.
2. An additional threat to loss of biodiversity knowledge is the drastic loss of the world's linguistic diversity. As languages are lost, so too is the vocabulary used to identify and describe "local flora, fauna, native traditions and knowledge" (Nettle, Daniel and Suzanne Romaine. *Vanishing Voices: The Extinction of the World's Languages.* Oxford University Press, 51).
3. In the decade between 1992 and 2002 servings of 8 oz cola drinks increased from 275 per person per annum to 487 per person per annum (GRAIN 2015). By the early 2000s over 90% of Coca-Cola and PepsiCo

sales were from *tiendas,* increasingly owned by junk food processors, such as OXXO, a subsidiary of Coca-Cola (GRAIN 2015).

4. Is important to note that the "introduction of European land tenure systems in Africa and Asia has often resulted in the usurpation of women's customary land rights and their transfer to men" (Howard 2003b, 16).

References

Altieri, Miguel. 2011. "Scaling Up Agroecological Approaches for Food Sovereignty in Latin America." In *Food Sovereignty: Reconnecting Food, Nature and Community,* edited by Hannah Wittman, Annette Aurelie Desmarais, and Nettie Wiebe, 120–133. Cape Town, Dakar, Nairobi, and Oxford: Pambazuka Press; Oakland: Food First Books; Bangalore: Books for Change; Winnipeg: Fernwood Publishers.

Barker, D. 2002. "Globalization and Industrial Agriculture." In *Fatal Harvest: The Tragedy of Industrial Agriculture*, edited by Andrew Kimbrell. Washington and London: Island Press.

Batal, Malek, and Elizabeth Hunter. 2007. "Traditional Lebanese Recipes Based on Wild Plants: An Answer to Diet Simplification?" *Food and Nutrition Bulletin* 28 (2) (June): S303–S311.

Blythman, Joanna. 2004. *Shopped: The Shocking Power of British Supermarkets.* London: Fourth Estate.

Carlsen, Laura. 2004. "The People of Corn." *New Internationalist*, December.

Critser, Greg. 2003. *Fat Land: How Americans Became the Fattest People in the World.* New York and London: Allen Lane.

CSM Working Group of Women. 2019. *Without Feminism There Is No Agroecology: Towards Healthy, Sustainable and Just Food Systems.* Edited by Teresa Maisano, Civil Society and Indigenous Peoples' Mechanism (CSM) for Relations with the UN Committee on World food Security (CFS) (August).

Davis, Mike. 2005. *The Monster at our Door: The Global Threat of the Avian Flu.* New York and London: The New Press.

Doss, Cheryl R., and Ruth Meinzen-Dick. 2020. *Land Tenure for Women: A Conceptual Framework*, October. https://doi.org/10.16/jlandusepol.2020.105080.

FAO. 2008. *Food Security: Concepts and Frameworks.* www.fao.org.

FAO. 2019. *The 10 Elements of Agroecology: Guiding the Transition to Sustainable Food and Agricultural Systems.* www.fao.org/agroecology.

Farming Matters. 2010. "Going for More Livestock: The Benefits of Livestock." ILEIA, March.

Friedman, Harriet. 1999. "Remaking 'Tradition': How We Eat, What We Eat and the Changing Political Economy of Food." In *Women Working in the*

NAFTA Food Chain: Women, Food and Globalization, Chapter 1, edited by Deborah Barndt. Toronto, ON: Second Story Press.

Gold, Kate Rory P. H., and Maria Ouma Mc Burney. 2012. *Conservation of Plant Diversity for Sustainable Diets.* Kew, UK: Royal Botanical Gardens; Canterbury, UK: University of Kent; Siaya Province, Kenya: TATRO Women's Group.

GRAIN. 2014. *Hungry for Land: small Farmers Feed the World with Less Than a Quarter of All Farmland*, 28 May. http://www.grain.org.

GRAIN. 2015. *Free Trade and Mexico's Junk Food Epidemic*, February. www.grain.org.

GRAIN. 2016. *The Great Climate Robbery: How the Food System Drives Climate Change and What We Can Do About It.* Barcelona: GRAIN; Pan Africa: Daraja Press; New Delhi: Tulika Books; Melbourne: Spinifex Press.

GRAIN. 2018. *Don't Get Fooled Again! Unmasking Two Decades of Lies About Golden Rice.* GRAIN, MASIPAG and Stop Golden Rice Network, 21 November 2018.

GRAIN. 2019. *Breaking the Silence: Harassment, Sexual Violence and Abuse Against Women in and Around Industrial Oil Palm and Rubber Plantations*, March. Cameroon: RADD; Gabon: Muyissi Environment; Liberia: Natural Resource Women Platform; Sierra: RADIO Culture; Leone: World Rainforest Movement (WRM).

GRAIN. 2020. *What Do FTA's Mean for African Women? A Critical Look into the African Continental Free Trade Area*, 8 July. https://grain.org/article/6496.

Guthman, Julie. 2011. *Weighing in: Obesity, Food Justice and the Limits of Capitalism.* Berkeley: University of California Press.

Guthman, Julie. 2014. *Agrarian Dreams: The Paradox of Organic Farming in California.* Berkeley: University of California Press.

Hauter, Wenonah. 2012. *Foodopoly: The Battle Over the Future of Food and Farming in America.* New York: The New Press.

Hirata Kimura, Aya. 2013. *Hidden Hunger: Gender and the Politics of Smarter Foods.* Ithaca and London: Cornell University Press.

Holt-Giménez, Eric. 2017. *A Foodie's Guide to Capitalism: Understanding the Political Economy of What We Eat.* New York: Monthly Review Press and Food First Books.

Holt-Giménez, Eric. 2019. *Can We Feed the World Without Destroying It?* Cambridge, UK: Polity Press.

Howard, Patricia L. 2003a. *The Major Importance of Minor Resources: Women and Plant Biodiversity* Gatekeeper Series, no. 112. International Institute for Environment and Development.

Howard, Patricia L. 2003b. "Women and the Plant World: An Exploration." In *Women and Plants: Gender Relations in Biodiversity Management and Conservation*, edited by Patricia L. Howard, 1–48. London and New York: Zed Books.

Howard, Patricia, L. 2006. *Gender Bias in Ethnobotany: Propositions and Evidence of a Distorted Science and promises of a Brighter Future*. Conference: Distinguished Economic Botanist Lecture, Kew Royal Botanical Gardens, November.

Kennedy, Gina, Dietmar Stoien, Danny Hunter, Enoch Kikulwe, 2017. "Food Biodiversity for Healthy, Diverse Diets." In *Mainstreaming Agrobiodiversity in Sustainable Food Systems*, 10–13. www.biodiversityinternational.org.

Kimbrell, Andrew. 2002. "Monoculture Versus Diversity: The Illusion of Choice." In *Fatal Harvest: The Tragedy of Industrial Agriculture*, edited by A. Kimbrell. Washington and London: Island Press.

Kloppenburg, Jack. 2011. "Seed Sovereignty: The Promise of Open Source Biology." In *Food Sovereignty: Reconnecting Food Nature and Community*, edited by Hannah Wittman, Annette Aurelie Desmarais, and Nettie Wiebe, 152–165. Cape Town: Pambazuka Press; Dakar and Nairobi: Oxford Books; Winnipeg: Fernwood Publishing; Oakland: Food First Books; Bangalore: Books for Change.

Korol, Claudia. 2016. *Land for the Women Who Work It: Struggles in Latin America*. Barcelona, Spain: GRAIN (October).

Lang, Tim and Michael Heasman. 2004. *Food Wars: The Global Battle for Mouths, Minds and Markets*. London: Earthscan.

LEISA Magazine. 2009. "Women and Food Sovereignty," September, 25 (3).

Mies, Maria, and Vandana Shiva. 1993. *Ecofeminsm*. Melbourne: Spinifex Press.

Montiel, Mart Soler, Marta Rivera-Ferre, and Irene Garcia Roces. 2020. "The Path to Feminist Agroecology." In *Agroecology and Feminism: Transforming Our Economy and Society*, 7–10. *Farming Matters*, October. www.agricultures network.org.

Mpofu, Elizabeth. 2018. "Agroecology Is Our Best Hope for Sustainable Development." *Farming Matters*, March, 33. www.agriculturesnetwork.org.

Murali, Vani Swarupa. 2020. *In India and Africa, Women Farmers Lack Land Rights*, June. https://www.lowyinstitute.org.

Nazarea, Virginia D. 2005. *Cultural Memory and Biodiversity*. Tuscon: The University of Arizona Press.

Nestle, Marion. 2002. *Food Politics: How the Food Industry Influences Nutrition and Health*. Berkeley and London: University of California Press.

Patel, Raj. 2011. "What Does Food Sovereignty Look Like?" In *Food Sovereignty: Reconnecting Food, Nature and Community*, edited by Hannah Wittman et al. 186–195. Cape Town, Dakar, Nairobi, and Oxford: Pambazuka Press.

Patel, Raj, and Jason M. Moore. 2018. *A History of the World in Seven Cheap Things: A Guide to Capitalism, Nature, and the Future of the Planet*. Melbourne: Black Inc.

Pearce, Fred. 2012. *The Land Grabbers: The New Fight Over Who Owns the Earth*. Boston: Beacon Press.

Pepper, D. 2006. "Indians Get a Taste of Supermarkets". *The Age*, Melbourne, November 4. https://www.theage.com.au/world/indians-get-a-taste-of-supermarkets-20061104-ge3hs8.html.

Peschard, Karine, and Shalini Randeria. 2020. "Keeping Seeds in Our Hands': The Rise of Seed Activism." *The Journal of Peasant Studies*. https://doi.org/10.1080/03066150.2020.1753705.

Pimbert, Michel. 2018. "Global Status of Agroecology: A Perspective on Current Practices, Potential and Challenges. *Economic and Political Weekly* 53 (41): 52–57. https://www.epw.in/journal/2018/4/1review-environment-and-development/globalstatus-agroecology/html/.

Pollan, Michael. 2006. *The Omnivore's Dilemma: The Search for a Perfect Meal in a Fast-Food World*. London: Bloomsbury.

Popkin, Barry M., Linda S. Adair, and Shu Wen Ng. 2012. "Now and Then: The Global Nutrition Transition: The Pandemic of Obesity in Developing Countries." https://www.ncbi.nim.nih.gov/pmc/articles/PMC3257829/.

Ross, Lois. "Land Grabbing a Growing Concern as Climate Change Threatens Food Production," 27 November. https://farmlandgrab.org/29346.

Rosset, Peter M. 2006. *Food Is Different: Why We Must Get the WTO Out of Agriculture*. London and New York: Zed Books.

Sachs, Carolyn. 1996. *Gendered Fields: Rural Women, Agriculture, and the Environment*. Boulder, CO: Westview Press.

Scrinis, Gyorgy. 2013. *Nutritionism: The Science and Politics of Dietary Advice*. Sydney, Melbourne, Auckland, and London: Allen and Unwin.

Seremetakis, C.N. 1994. *The Senses Still: Perception and Memory as Material Culture in Modernity*. Chicago: University of Chicago Press.

Schlosser, Eric. 2001. *Fast Food Nation: What the All-American Meal is Doing to the World*. New York and London: Allen Lane.

Shepard, Daniel and Anuradha Mittal. 2009. *The Great Land Grab: Rush for World's Farmland Threatens Food Security for the Poor*. Oakland, CA: The Oakland Institute.

Shiva, Vanadana. 2000. *Stolen Harvest: The Hijacking of the Global Food Supply*. Cambridge, MA: South End Press.

Teuteberg, H. J. 1975. "The General Relationship Between Diet and Industrialization." In *European Diet from Pre-Industrial to Modern Times*, edited by Elborg and Robert Forster, 61–109. London and New York: Harper Row.

Tracey, Mollie. 2020. "US Investor Snaps Up Almond Orchard for $98 Million." Farmlandgrab.org, 6 September. https://farmlandgrab.org/129846.

Tutwiler, Ann M., Arwen Bailey, Simon Attwood, Roseline Remans, and Marleni Ramirez. 2017. *Agricultural Biodiversity and Food Systems Sustainability*. Biodiversity International. Mainstreaming Agrobiodiversity in Sustainable Food Systems: Scientific Foundations for an Agrobiodiversity Index.

UN Women. 2017a. "Rural Women Tackle Drought-Affected Mozambique's Rise in Child Marriage." https://www.unwomen.org/en/news/stories/2017/11/feature.

UN Women. 2017b. "Insecure Land Rights for Women Threaten Progress on Gender Equality and Sustainable Development," July.United Nations Human Rights Special Procedures. https://unwomen.org.

Vispo, Isabel Alvarez, and Ingrid Paola Romero Nino. 2020. "Can Feminist Agroecology Be Scaled Up and Out?" In *Agroecology and Feminism: Transforming Our Economy and Society.Farming Matters*, October. www.agricultu resnetwork.org.

Wahlquist, Mark L. 2003. "Regional Food Diversity and Human Health." *Asia Pacific Journal of Clinical Nutrition* 12 (3): 304–308.

Wallace, Rob. 2016. "Did Neoliberalizing West African Forests Produce a New Niche for Ebola?" In *Big Farms Make Big Flu: Dispatches on Infectious Disease, Agribusiness, and the Nature of Science*, 324–335. New York: Monthly Review Press.

Waring, Marilyn. 2004. *Counting for Nothing: What Men Value and What Women are Worth*. University of Toronto Press.

Wilson, Bee. 2019. *The Way We Eat Now: Strategies for Eating in a World of Change*. London: 4th Estate.

Wise, Timothy A. 2019. *Eating Tomorrow: Agribusiness, Family Farmers, and the Battle for the Future of Food*. New York: The New Press.

CHAPTER 8

Women Feed the World: Biodiversity and Culinary Diversity/Food Security and Food Sovereignty

For millennia throughout the world women have been the preparers, cooks and conservers of food, as well as the primary gatherers of wild edible plants and growers of much of the plants eaten by their families and communities. Indeed, it is likely that women as food-gatherers developed early agriculture, deliberately cultivating wild grains that they had gathered from the hills since 'time immemorial' in the Fertile Crescent (Fernand Braudel 2002, 40). Reports from early European settlers in Australia described how Aboriginal women in their role as food-gatherers did exactly this: "From their woven dilly bags (they) sprinkled seed food over the ground with repeated applications of water to make them grow" (cited in Pascoe 2014, 30).

Because indigenous, rural and peasant women have had to depend on what was locally available, and in their role as food preparers and cooks for their families, they have, over many millennia, developed ways of preparing, preserving and detoxifying plants to make them safe, nutritious and palatable. This has involved ingenuity, keen observation and memory, the knowledge and skills involved in this taking at least one-third of a lifetime to learn (Howard 2003a). The constraints of having to make use of what is locally available has resulted in a reciprocal relationship between food preparation and local biodiversity, both of which women have traditionally been the custodians. Women have grown, harvested and gathered local plants, both wild and cultivated, devising countless ways

V. A. Swinbank, *Women's Food Matters*,
https://doi.org/10.1007/978-3-030-70396-7_8

of preparing meals from them, the cumulative effect being the creation of distinct regional cuisines throughout the world. Indeed, it would not be overstating it to say that without women's food knowledge and skills over countless centuries, it is doubtful that humanity would have survived, let alone thrived. Yet this is a fact that is scarcely acknowledged or recognised, except by a few feminist researchers and scholars, principally those in the field of ethnobotany.

Ethnobotanist Patricia Howard has found from her extensive research over many years that, despite the fact that over the last couple of decades ethnobotany research has greatly increased, corresponding research in women's role and acknowledgement of women's role in developing and maintaining biodiversity has increased only very slightly (Howard 2006). Howard points out that one of the main reasons for this gender bias is the fact that "because most plant use, management and conservation occurs within the domestic realm, and because the principal values of plant genetic resources are localised and non-monetary, they are largely invisible to outsiders and are easily undervalued" (Howard 2003a, 2). However, the relationship between biodiversity and women's plant and food knowledge is an area that is becoming of increasing interest to ethnobotanists, ethnobotany being the "study of plant-people relationships" (2003b). Carine Pionetti (2005), from her detailed research on women farmers in dryland India arrives at a similar conclusion: "While most ethnobotanical research tends to overlook gender as an important variable, gendered approaches to the study of people-plant relations leave no doubt about the inter-relationship between plant biodiversity, gender roles, household food security, culinary knowledge and cultural practices" (85). On a global basis, and especially in the Global South, women are the principal wild plant gatherers, and cultivators of homegardens in which they grow food for the family and herbs for culinary and medicinal purposes. They also play a major role in agriculture, in which they are also the principal seed custodians. Nevertheless, women's traditional plant varieties are considered, particularly by outsiders, as 'minor' resources, with women being 'minor' actors, "secondary to men who are presumed to be the knowledge holders, managers and preservers of plant resources such as staple crops and trees that are considered 'valuable'" (2005, 3). There are a number of reasons for this lack of recognition, such as the masculinist tradition of overlooking women's achievements, whilst claiming them as their own. In addition women generally do not 'blow their own trumpet' in the way that men often do; they quietly go about their busy days of looking after others, with their food preparation and cooking being taken for granted by others and even by themselves. As Indian ecofeminist Vandana Shiva says: "Women's work in food and agriculture has been made invisible, even though it is the foundation of society" (2016, 113).

Throughout human history women have been the principal providers and preparers of food. According to FAO estimates, women produce more than 50% of the world's food, and in much of the non-industrialised Global South peasant women still produce 60–80% of food for local consumption (GRAIN 2016, 85), as well as being the custodians of agricultural and horticultural biodiversity. However, this is increasingly under threat from modern industrialised agriculture which is falsely promoted as the solution to world hunger, when in fact the real solution lies in protecting and supporting the sustainable and biodiverse food systems that have been managed for millennia by small-scale peasant farmers, mainly by women. In reality "only 30 percent of the food that people eat comes from large-scale industrial farms. The other 70 percent comes from small-scale farmers" (Pimbert 2018. 55). As I discussed in the previous chapter, not only is industrialised agriculture not the solution to world hunger, but it in fact causes massive environmental damage and is a major contributor—between 44 and 57%—to climate change due to its total dependence on fossil fuels (GRAIN 2016).

Of course, traditionally both men and women are involved in agriculture. Men's role tends to be in grain production, especially that which is grown for the market or for export, although even in this domain women often contribute the majority of the labour. Women's crops, on the other hand, are typically vegetable or root, with subsistence food grown for local consumption (Sachs 1996, 68). Also, in contrast to men's focus on one or two staple crops, in traditional agricultural systems, women employ multicropping strategies, "rarely organising their lives around production of one particular crop, instead engaging in multiple productive and reproductive activities" (73). The reason for this is that women's food production is intimately and inextricably linked to their role as preparers and cooks for family sustenance and nutrition. Consequently, they select, save and propagate seeds of plants according to their culinary and medicinal applications, ensuring that they have a regular and varied supply of plants according to seasons and growing conditions. As Carolyn Sachs expresses it, the closer the food is to the table the more likely women are to do the work associated with it (1996, 2). In addition to plant selection and cultivation, women are also responsible for post-harvest processing and family food supplies which "means that they try to ensure that varieties are in line with culinary traditions, are palatable and nutritious, and meet processing and storage requirements" (Howard 2003a, 10). There are several reasons for women's greater knowledge of plants, including their historical role as gatherers, plus the application of plants, both cultivated and wild, to culinary preparation, as well as to other subsistence needs. Consequently, throughout much of the world women traditionally manage the majority of useful plant species (Howard 2003a, 2003b).

The Importance of Women's Homegardens

Over countless generations women have gathered wild plants for food and medicinal purposes, saving their seeds and selecting the best for propagation in their homegardens. They are, in fact, "the oldest and most widely used cultivation systems on the planet" (Howard 2003a, 7). Globally, homegardens are the most widespread land use system, "living evidence of the biological and social-cultural interactions that develop around local livelihoods and people's efforts towards well-being" (Lopez-Alzina and Howard 2012, 17). Women's homegardens are also a major repository of plant biodiversity, having a "far greater species diversity than cultivated fields, and hence should be recognised as the single most important repository of plant diversity" (Howard 2003a, 7). However, because women's work is defined as being in the domestic realm, women's work and knowledge of plants are generally considered unimportant, even invisible, with women's work and knowledge in biodiversity conservation often being disparaged (especially by Western observers) as 'minor' or 'supplemental' to agricultural production (8). Because women's work is subsistence and unpaid, and even though it makes the major contribution to household sustenance and well-being, it is considered to be less important than men's involvement in cash crops and paid labour. As Fatima Osman Ibnouf (2009) asserts, speaking of women's central role in household food security in Sudan: "If women's unpaid work was properly valued they would emerge in most societies as the major breadwinners given their greater contribution of working hours than men" (144).

Homegardens are usually situated close to the house or home compound so that they are easily accessible for food preparation and other domestic purposes. They are also "organised by the rules of the kitchen in order to have fresh and diverse food all year round" (Tittonell

2015, 9). The majority of studies on homegardens are those of the tropical and subtropical regions of the world, especially Africa, Asia and Central America. Homegardens in these regions consist of a "multi-strata agroforestry system....that is, characterized by a set of land use practices involving deliberate management of multipurpose trees and shrubs in intimate association with annual and perennial agricultural crops and invariably livestock" (9). They are also characterised by traditional biodiverse methods of cultivation, using organic fertilisers, such as compost and manure, natural pest control, crop rotation and polyculture. In addition to supplying diverse, year-round healthy food for the family, agroecological systems are inexpensive due to their minimal dependence on external inputs. All of these features are of particular importance to women.

Not surprisingly, there is an intimate relationship between women's homegardens and their culinary activities, each reflecting and reinforcing the other. However, most ethnographic studies neglect the essential role of cuisine in its relationship with biodiversity, viewing both areas as distinct and relatively unconnected. Patricia Howard (2003b) challenges this view, saying that: "The kitchen is quite possibly the most undervalued site of plant biodiversity and conservation... Only rarely is it considered that women's work in the kitchen is essential to biodiversity conservation" (8, 13). This is because most food-related ethnobotanical research focuses on men's role in agriculture, overlooking women's vital role in both food production and food preparation. Nevertheless, the protection of plant genetic diversity depends upon an understanding of its inextricable connection to regional cuisine, with the study of cuisine involving "food selection prior to plant production or procurement and extends through plant cultivation and management, harvest, food preparation and consumption" (63).

Homegardens are of great importance to women who migrate to new places, either by choice or necessity. As soon as they are settled, if they have the space, women will often establish a home plot in which to grow their traditional crops as well as perhaps incorporating some new ones indigenous to their new environment. Laurie Greenberg (2003) describes how Yucatan women attempt to preserve their culinary traditions when they migrate to the Caribbean coast of Mexico. Home gardens are crucial to this, in which they grow up to 140 plant species, consisting of vegetables, herbs and fruit. *Limon* (Mexican or West Indian lime) and sour (Seville) orange are used for flavouring soups, stews and fish, and condiment pastes, "consisting of herbs and spices ground together in specific proportions and dissolved in lime or sour orange juice to produce distinctly Yucatec flavours" (60). Growing plants that are essential to their cuisine "enables women to continue to provide foods that conserve aspects of Yucatec cuisine and thus Yucatec ethnicity" (52), as well as their house-lots being sites of in situ plant preservation. Both the ingredients and cooking techniques of Yucatec cuisine have their origins in Mexico, Spain and the Caribbean, which shows how women, far from being rigidly conservative in their cooking—an attitude expressed by Revel et al. and which I refute in Chapter 3—are inquisitive experimenters and culinary innovators, exploring new ingredients and experimenting with new cultigens in their small plots. As Greenberg explains, Yacatec women pass on their knowledge intergenerationally, and "under women's management, some plants evolve, are conserved and are disseminated throughout the world" (51). Indeed, the "interrelationships between women, the maintenance of culinary traditions, cultural identity and biological diversity are ...made very evident in the exchange and movement of plant genetic resources through migration" (Howard 2003b, 11). In traditional Yucatec households, women's daily activities in the house lot "include horticulture, small livestock production, child rearing, food processing and preparation and related tasks" (Greenberg 2003, 51). In addition to being vital repositories of genetic diversity, homegardens in many cultures serve an important function for children to learn about natural and social relations: "the repeated practice of tasks, celebrations and everyday activities, such as those associated with women's gardening and cuisine, that result in an unconscious conditioning throughout childhood and a way of operating in the world" (53). Ethnobotanist Shirley Hoffman (2003) explains that women take great pride in their homegardens, with experienced cultivators enjoying having multiple crops and varieties in their gardens and fields in order to achieve a broad mix of different varieties of plants. Women's plant knowledge and skills in food production also confer considerable social status on them. Sadly, however, this is being eroded with the incursion of modern education and technology (263). Homegardens also have an aesthetic value for women, who "will put in a considerable amount of extra work for little or no return simply for the pleasure of trying something new or having an attractive plant" (263).

Small-scale peasant farming and homegardens have always been repositories of biodiversity and experimentation. Indeed, "peasant agriculturalists everywhere in the world have proven to be quick to adopt new plants and new techniques to improve agricultural production" (Handy and Fehr 2011, 48). Contrary to what might be commonly thought, it was not large farms that "fostered the spread of New World agricultural products around the world from Ireland to China", but small-scale agriculturalists, "as the small size of each person's rows allowed for small changes in agricultural techniques or new crops while others in the region watched to see how the experiment turned out" (48). This was/is facilitated by the plots being close to the household, and therefore being able to be closely observed and tended on a daily basis. As a result of this process, "a host of new cultigens emerged from the Americas and remarkably quickly became peasant crops in locales all around the world" (48).

The transatlantic transmission of food crops from the New World to Europe and beyond also occurred in the opposite direction, that is, from Africa to the New World, one example being rice. Although it is generally assumed that rice cultivation in America is of Asian origin, it was in fact African slaves who were instrumental in introducing rice from West Africa to South Carolina, with female slaves bringing with them their traditional knowledge of the many varieties of rice native to West Africa. "Throughout the West African rice region seed selection is the responsibility of females", where rice culture which embodies "specialized knowledge systems, is often the domain of African women" (Carney 2001, 49, 107). In addition to rice, West African slaves to Carolina introduced other African crops such as okra, various greens, millet, sorghum, and black-eyed peas which they grew in their 'provision gardens'. The introduction of African crops, in turn, "profoundly affected foodways and culinary traditions in areas of the Americas settled by blacks" (156). As James C. McCann explains:

> The new infusions of West African cultures in large numbers, especially via women, brought generations of experience and ideas about the preparation of food from African culinary practice, which had adapted to include the numerous New World crops …of staples like cassava, cocoyam, maize and capsicum peppers…The results formed the foundation of signature dishes that have become symbols of identity and culinary traditions in North America and the Caribbean rim….., such as Cuban *ajiao*, Caribbean rice and beans and New Orleans jambalaya. (McCann 2010, 165–166)

'Women Have Always Gathered Wild Plants': Women as Food-Gatherers

Human beings have since time immemorial foraged for wild plants for food, fibre, medicinal purposes and fuel, as well as materials for construction. This was the case in pre-agricultural hunter-gatherer societies, as it still is in those few such societies that remain (Sachs 1996). Many agricultural societies continue to depend on wild plants for nutrition and dietary diversity, in addition to fuel and fodder. Indeed, globally about one billion people consume wild foods on a regular basis (Howard 2012, 117). However, because field crops of grain, such as wheat and rice, provide the food staples of agricultural societies, agricultural researchers and modern diet specialists "tend to ignore wild plants compared to dietary staples in part because of a lack of understanding about which plants are used and the extent to which they contribute to routine cuisine" (Ertug 2009, 64). This ignorance is also greatly multiplied because it is women who are the principal players in wild plant gathering, and management (Howard 2003a). Because wild plants tend to be associated with poverty and lack of other sources of nutrition, their central importance in cuisine is generally overlooked. Wild or uncultivated plants, nevertheless, often play a major and essential role in many rural regional cuisines. For instance, in a study carried out in Bangladesh it was found that: "Far from being a minor supplement or simply a food crisis food, the leafy greens, tubers and small fish collected by people from the lands and water bodies of their communities are a vital part of their daily diet" (Mazhar et al. 2007, 40). Significantly, this study also found that such food was highly valued by farming families, regardless of social class. This is because these foods are of great importance to local culinary culture due to their particular flavours, as well as being highly nutritious. Similarly, in Nepal women also are the primary holders of indigenous knowledge of *jangal* (wild and medicinal foods) resources, which constitute an important part of the diet, especially when field crops are dwindling, thereby ensuring that food is available throughout the year (Daniggelis 2003).

Turkish ethno-archaeologist and ethnobotantist, Fusun Ertug refutes the common belief that wild plants are relied upon by rural communities mainly in times of famine, arguing that, even in times of plenty, women continue to value them for their nutritive and medicinal values, as well as their taste, which is an important part of their culinary cultural identity. In research carried out in rural Anatolia and the Turkish Aegean region it was found that up to 300 wild plants were still being used, of which 100 were used as food (2009, 65), with the numbers used varying regionally, depending on the degree of urbanisation, where such use tended to be less. Moreover, rural people who have migrated to towns and cities still value wild plants for their taste, nutritional and medicinal values, which they buy from women selling them at markets. In addition to their taste, wild plants, being rich in micronutrients, vitamins and minerals are highly valued for their important role in complementing staple foods (Dogan 2012). The main plants gathered by women are leafy greens, especially when they are at their best during winter and spring, contributing important vitamins and minerals to the diet (Ertug 2003, 189). The gathering of leafy greens is traditionally done by women in Anatolia (up to 40 different varieties), as it is by women throughout the world. Children, especially daughters, accompany women, both to help and to learn about wild plants and their uses. Italian food writer, Valentina Oliveri, describes how, as a child, she would accompany her Calabrian grandmother who "knew all the herbs that grew wild in that ancient land....the knowledge, involving both body and mind, being passed on quite naturally...as women have been doing for thousands of years" (2015, 98). Similarly, Patience Gray (1987), in her classic book '*Honey from a Weed*', which describes her many years of living off the land in Tuscany, Catalonia, the Cyclades and Apulia, witnessed a little seven-year-old girl gathering weeds from her father's vineyard in Tuscany. Gray was amazed at the extent of the vocabulary of the little girl, who, as she gathered, said: "'This is for cooking' or 'This is for salads'" (190). Fusun Ertug explains that rural Turkish men also gather, although far fewer species than women, the main one being mushrooms. Women's and men's approach to gathering is also very different; whereas women prefer to gather in social groups sharing some of the produce with friends and relatives, men tend to work alone, especially when they gather mushrooms, as they "don't want to reveal their sources to others" (2003, 187). In addition, unlike women, they do not give what they gather as gifts, keeping it instead for their own consumption. For women, food-gathering not only provides valuable food for their families, but is also valued as a social activity, enabling them to get out of the house and socialise with other women, building networks and sharing skills and knowledge (Ertug 2003).

'What Do You Mean by Weeds?': Culinary Applications of Wild Plants

Significantly, "knowledge concerning wild foods is more likely to survive since it is deeply entrenched in local cuisines, traditional tastes and women's social organisation" (Ertug 2003, 194). This assessment is confirmed by ethnobotanists, Andrea Pieroni and Nello Biscotti (2015), who, based on their research in northern Apulia in SE Italy, found that the use of wild plants "in folk cuisines is still very much alive", despite changes in the traditional lifestyles of the peasants. This is because of the "resilience of a plant-based cuisine which retains a crucial cultural value in northern Apulia and in SE Italy in general" (325, 327). Symbolic of wild vegetable-based cuisine in Apulia is an omelette made with a wild edible bulb. In addition there exist dozens of dishes based on the use of leaves, bulbs, roots, shoots, fungi, fruits and flowers of dozens of wild plants which are used in raw and cooked salads, soups, with pasta, with eggs, in pies, roasted with meat, fried, pickled and eaten raw as a snack. Similar culinary use of wild plants is common throughout much of the Mediterranean region, especially those areas that are less urbanised. Elsewhere, ethnobotanical research in Albania has found unripe wild fruits being used as yoghurt starters, as well as for making vinegar, and the use of wild apples and grapes as yeast starters for bread-making (Pieroni and Soukand 2017).

In her book of the cuisine of the Greek island of Ikaria, Diane Kochilas (2014) describes the culinary uses of some of the "hundreds of edible and potable wild plants that carpet Ikaria", which traditionally have been a major component of the local cuisine, and which contribute to the legendary longevity of the Ikarian population. Wild plants are used on their own or with other wild plants and/or cultivated vegetables, in combination with grains and pulses, dairy, fish and meat. Some of these plants include: wild amaranth; wild asparagus, which is cooked in omelettes or blanched and dressed with olive oil and lemon juice; blue mallow, used in savoury pies; borage and wild chervil, used in salads and pies; wild fennel, used in countless dishes; fiddle-head ferns, fried with eggs or pickled; lemon balm, and Mediterranean hartwort, used in savoury pies; nettles used in soups and vegetable dishes; purslane, in salads and cooked with garlic and yoghurt; taro leaves; yellow salsify, eaten raw and cooked and pickled; bitter dock, used in savoury pies; black bryony, cooked with eggs; chicory, boiled for salads: dandelion, eaten in salads and cooked; golden thistle, Mediterranean mustard, white mustard, prickly goldenfleece, shepherd's purse, all boiled for use in salads; the list goes on! (50–52). In addition to this are wild herbs, used for culinary and medicinal purposes, such as many kinds of mint, oregano, savory, thyme, sage, rosemary, rue, elder, hawthorn and St. John's wort.

Patience Gray, referring to her many years of living in Greece and Italy, says that her interest in edible weeds was inspired by seeing women and children on the Greek island of Naxos gathering wild plants in early Spring. The plants were referred to as *radikia*, indicating their beneficial roots, as well as their leaves (*horta*). When gathering, only a part of the root, which is valued for its bitter medicinal properties, is taken, so that the plant can continue to grow. The bitterness of many wild Spring plants is particularly valued as a liver cleanser. As on Ikaria, the variety of edible wild plants on Naxos is huge: dandelions, wild chicory, endive and salsify, ox-eye daisy or marguerite, thistles, wild carrot, wild fennel and chervil, wild mustard, various rockets, pennyroyal, wild thymes and savory and more (Gray 1987, 188). In Apulia, Gray observed the gathering by women of bulbs such as tassel hyacinth which are "delicious in a salad with olive oil and wine vinegar". She points out that many wild weeds are either "cooked or served as a salad,....employing a variety of different plants" (194) both for flavour and maximum nutritive value.

The Balkans is also a region with a great wealth of wild plant varieties that are incorporated into the local regional cuisines. For instance, an early-Spring dish of nettles, which are believed to purify the blood, is prepared in Bulgaria and Macedonia consisting of pureed cooked nettles, unsalted butter, ground walnuts and thick yoghurt. Another characteristic Turkish/Bulgarian/Macedonian dish uses edible wild greens such as fat hen and purslane baked with eggs, yoghurt and feta cheese (Kaneva-Johnson 1999). Turkey, due to its climate and geographic position, is particularly rich in wild plant diversity, the traditional use of which is "reflected in the rich Turkish cuisine" (Dogan 2012). The list of wild plants and their culinary application is vast, with wild plants used in an immense variety of dishes—soups, salads, pies and stews.

What are often dismissed as weeds by modern industrial agriculture, and which as a result are targeted with toxic herbicides, such as Monsanto's notorious glyphosate Roundup, are highly valued as nutritious and tasty food by many rural communities throughout the world. This is expressed by a female farmer from the Deccan region of India when she says: "What do you mean by weeds? There is nothing like a weed in our agriculture. In our agriculture all plants are edible" (Mazhar et al. 2007, 18). In traditional rural societies, such as that described above, "the entire social landscape is host to plants, both cultivated and uncultivated. And all of them are edible. They are food for humans or animals" (18). For these farmers there is no conflict or contradiction between cultivated and uncultivated plants; instead they are regarded as 'partner plants' growing in association with each other, with both being considered vital to a healthy and varied diet, as well as for other household needs (2007). For instance, wild plants and grasses growing in and alongside rice fields in Bangladesh are collected as fodder for farm animals; this is especially important in a situation such as Bangladesh where there is very little land for grazing. The use of herbicides makes mixed cropping impossible (18) as well as destroying uncultivated plants used for animal fodder, making the keeping of cows and goats much more difficult, especially for poor women. It also means that these nutritious 'weeds' that are rich in vitamins and minerals are no longer available, leading to malnutrition as well as impoverishing the local cuisine in which they are a vital component. Women farmers are vehemently opposed to the use of chemical pesticides and herbicides and take the lead in calling for the end of their use because of the harmful impact on themselves and their children and because of their destructive effect on valuable wild plants and on fish in nearby waterways (2007, 3). Their protest is also fuelled by the high number of male farmers who, indebted to chemical companies, commit suicide often by drinking herbicides.

The collecting of wild or uncultivated plants is just the first step in the culinary process. Women know what plants are suitable for particular dishes as well as their nutritive and medicinal properties, with bitter plants having medicinal values and sweet and sour special food values. In one Bangladeshi district it was found that 79 varieties of uncultivated leafy green vegetables were used by the local farming women, with many of the plants having very high nutritive value (34). In Bangladesh a common preparation is *bhaji*, consisting of freshly picked greens, onion, garlic and chilli fried in mustard oil. *Charchari* is a dry dish of greens, small fish, spices such as coriander, chilli paste and turmeric which combines well with partially fermented fish (31). *Ghonto* consists of cultivated vegetables, such as eggplant and potatoes mixed with uncultivated greens. Often several different greens are cooked "as a mixture in very precise combinations and in association with particular cooking styles" according to the seasons (32). The greens are also cooked with pulses, such as lentils, cowpeas and mung beans in curries or soups, accompanied by sorghum chapatti or one of several different varieties of millet, plus a hot sauce made from hibiscus pods or a condiment of ground sesame seeds. Such a tasty and nutritious combination means that "while the cultivated landscape provides carbohydrates and proteins in the form of grains, pulses and oil seeds, the uncultivated landscape completes the diet with the micro-nutrients contained in the greens" (34). The dependence on uncultivated plants is particularly important during the monsoon season when much of the land is flooded or being used for rice cultivation. This is because these plants provide essential micronutrients, such as iron, iodine and vitamin A, which are needed by young children and women who are pregnant or menstruating (47). Access to uncultivated food, as well as fuel for cooking, is vital for the survival of the rural poor; this is why it is crucially important that access to the commons is protected, an issue that I have addressed in Chapter 6. Gathering wild and uncultivated plants and other freely available food items is important to women "not only for economic reasons, but also because of a strong reluctance to lose their independence and identity" (Pionetti 2005, 72). Many of the traditional cuisines of the African continent are also reliant on the gathering of wild plants and greens. For instance, on "the savannas of West Africa alone, more than two hundred plants are still foraged for their edible fruits, seeds, and leaves" (Carney and Rosomoff 2009, 16).

'Agriculture Begins with Seeds and Ends on the Plate': Women as Seed Custodians

In rural societies throughout much of the world women are not only the main gatherers of wild or uncultivated plants, but are also, in their role as farmers, the principal custodians of agricultural seeds. While both male and female farmers select and grow crop varieties, women generally grow far more varieties than men; this is because, in their role as food providers and preparers, they are keenly aware of the need for crop variety for household food security (Pionetti 2005, 92). Carine Pionetti (2005) also points out: "Research findings from different parts of the world suggest that women and men often hold a different set of perceptions and knowledge about crop varieties grown on their land...(with) women very often us(ing) a broader set of selection criterion than men" (94). Women's crop variety criteria are based on considerations to do with production, processing, storage and preservation in addition to nutritional and culinary qualities. From her research in semi-arid Andhra Pradesh, Pionetti found that women farmers want most of the family's daily diet to come from their own land. This is not only for economic reasons, as well as household food security, but also because they consider that their own pulses and cereals have greater variety and better flavour than shop-bought ones. Similarly, based on her research in the Philippines, Virginia Nazerea (2005) found that "Women in general understand the intricacies of the production and marketing of rice, vegetables, and ornamental plants and utilize more criteria than men, ranging from agronomic to gastronomic, in recognizing and categorizing crop varieties" (86). Women's knowledge of plant varieties and of food crop germplasm is a consequence of their central role in subsistence production. This is because culturally they "have the responsibility of seeing that everyone in the household gets fed and cared for on a day-to-day basis" (81). Consequently, in subsistence societies, it is women who hold the key to the preservation of both biodiversity and the cultural memory that it supports, and who act as "conduits of genetic and cultural transmission" (81).

Rural women's intimate and detailed knowledge of plant genetics and seed selection is repeated throughout most of the Global South, in Asia, Africa and Latin America. However, when modern Western agricultural methods and cash crops are introduced, women's experience and knowledge about agricultural sustainability are overlooked or are dismissed as backward. The introduction of chemically-based cash crops and monocropping, which focuses on male producers at the expense of women, undermines food security and depletes soil fertility. This stands in contrast to women's sustainable approach to agriculture whereby they grow a wide range of food crops to ensure a stable food supply which protects agricultural biodiversity and also helps protect against the risks of climate variability. For instance, in the dryland Deccan region of India, in order to minimise the risk of crop failure, women farmers grow a variety of grains and pulses, including foxtail millet, sorghum, field beans, cowpeas, pigeon peas and sesame. For Dalit women the cultivation of yellow sorghum is important, despite its low market value (and therefore is discouraged by agricultural scientists), because it grows in dry conditions and is highly nutritious. It also provides fodder for animals, as well as being able to be used in thatching and fencing. Yellow sorghum is grown as a companion crop to pigeon peas, which in turn are prepared in dishes to be eaten together, thereby increasing their nutritive value. The complimentary connection between agriculture and cooking highlights the fact that the crops that women grow are:

> indicative of their food culture, and the relationships between foods in the kitchen reflect relationships in the field…This unique 'farm-to-kitchen' model is what has kept agrobiodiversity alive on their farms for centuries. Since women are also the most important torchbearers of this food tradition, they are also carriers of the agrobiodiversity. (Satheesh 2014, 42)

Women's priorities, in addition to providing a reliable and varied supply of food for their families, include fodder for livestock, fibre and any surplus for sale (Pionetti 2005, 91). Biodiversity and agroecology is especially important to poor, vulnerable women. Women employ agroecological methods using crop rotation to protect against pests and diseases, and build up soil fertility with organic matter. Female peasant farmers' interest in what they grow is driven by their household's nutritional needs and food security, rather than growing crops for commercial gain. Women farmers feel so strongly about the need to grow nutritious food crops, that when men's cash crops, such as cotton, become the driving force and consume much arable land, women resort to strategies to retain control over grains and seeds, such as concealing information from their husbands, and planting their precious food seeds surreptitiously (Pionetti 2005, 146). The priority rural women give to producing their own food for the family is cross-cultural; Carolyn Sachs, for example, describes how historically on "US farms, women persisted in producing food for their families long after men redirected their efforts towards commercial crops" (1996, 71).

Women are responsible for making sure that the seeds they have saved and stored are preserved for the next season's planting to ensure their families' future food security; indeed, "so important is seed to women farmers that they will go hungry rather than eat them" (Pionetti 2005, 144). The importance of seeds to women is reflected in their mantra 'save seeds, no matter what', a mantra passed on from one generation to the next. Women farmers do not buy or sell seeds, but exchange them with others. Seeds are, in fact:

> one of the rare resources over which women farmers exert direct and uncontested control. Seed stocks are a bargaining chip for women to use in intra-household negotiations over land use...In the absence of direct control over property, land, capital and money – something that should be relentlessly questioned – it is essential that women retain their control over food grain and seed stocks. (144)

Women's preservation and maintenance of seeds involves an enormous range of skills and knowledge which are "an intrinsic part of their identity and self-esteem" (134). These skills and knowledge, along with the culinary traditions associated with them are passed on intergenerationally from mother to daughter. Women's role in saving and exchanging seeds with other farmers safeguards the agricultural biodiversity that is the basis of their livelihoods and culinary culture. Women's intimate involvement in the local seed economy "strengthens the transmission from mothers to daughter of skills in seed production, grain processing and food preparation. Through teaching and learning, knowledge, skills and technological change adapt and evolve" (156).

The variety of seeds and landraces cultivated by women farmers throughout the world is vast; for instance, in Rwanda women produce more than 600 varieties of beans (Howard 2003b, 10). Countless other examples represent the incredibly rich biological cultural heritage of humanity, built up over many millennia, principally by women, for the reasons already explained. Their protection and preservation are also essential for human survival, especially in the current era of man-made climate change. But this vast wealth is under threat from the disastrous consequences of modern industrial agriculture, which itself is a major contributor to climate change.

A typical example of women farmers' biodiverse, agroecological approach to farming is that of Susilamma who, on her one-acre farm in the Deccan region of Andhra Pradesh, grows a mixture of cereals, legumes and oil seeds, including three varieties of sorghum, two varieties of foxtail millet, finger millet, little millet, pearl millet, kodo millet, three varieties of pigeon pea, two varieties of dollichos lablab, horsegram, cowpea, sesame, niger and two varieties of hibiscus (Mazhar et al. 2007, 58). These, plus the vast variety of uncultivated greens she harvests, especially during the monsoon season, "provide her with all the elements for a nutritionally balanced food basket, as well as plants of medicinal value" (58). All of her seeds come from those she has saved from the previous harvest; she saves the best seeds and stores them in a local palm leaf basket, mixing them with cow manure ash, and layered with neem leaves as protection against insects, so that they will "stay uninfested and viable for at least three years" (58). Susilamma farms without chemicals, with the soil fertility being built up with the manure from the farmyard, plus composted tree leaves and household waste. This fertility is reinforced by growing pigeon peas next to sorghum, which fixes atmospheric nitrogen on its roots, as well as providing leaf fall as green manure. The many trees that border her farm and that create boundaries within it enrich the soil and produce valuable biomass, as well as providing habitat for birds that feed on insects. This, plus the large number of beneficial insects due to the absence of pesticides, means that insect damage is minimal (58).

A classic example of women farmers' ingenuity in intercropping is that of Native American women in North, Central and South America in a system referred to as the 'three sisters' or *milpas*, in which maize, beans and squash are grown alongside each other (Sachs 1996). The first breeders of maize were Native American women, who initially cultivated it from a variety of wild grasses—teosinte—over the centuries, improving and selecting different maize varieties. In this multicropping system maize and beans are planted in hills, with the climbing beans supported by the maize, and with weed control and moisture retention provided by squash planted between the hills. In addition to making efficient use of growing space, this multicropping system helps prevent plant disease and pests, as well as the beans providing nitrogen for the maize. The combination of beans and maize is also nutritionally important as they provide a desirable protein balance (76). As Carolyn Sachs points out, modern science confirms that the "maize-beans-squash triad is ecologically, nutritionally and economically sound" (76).

Post-harvest Processing and Preservation

Historically, human beings have invented innumerable ingenious ways of preserving fresh foods, using methods that include salting, drying, smoking, pickling and fermenting. Throughout pre-industrial history women have been the main processors and preservers of post-harvest foods. There are countless examples throughout the world of women's ingenious methods of processing, detoxifying when necessary, and preserving, of which just a few examples are given here. These post-harvest activities are important for a number of reasons: they preserve the harvest, so that essential nutrients are available through the lean, non-growing months; they make potentially toxic plants, such as manioc, safe to eat; and they give cuisines distinctive flavours that form part of different culinary cultures. What was initially a need to preserve food, to deal with a surplus, to prevent it decaying and/or to preserve for it for winter, became in turn a vital ingredient in the local cuisine, valued for its taste, as well as its nutritional qualities. Furthermore, the "knowledge and skills required in the post-harvest food chain are complex and dynamic (with) several studies showing that traditional women's knowledge in food processing and storage often correlates with scientific knowledge" (Howard 2003a, 6). These post-harvest food practices are often done communally; women work together in groups, which makes for efficiency, sharing the work, skills and knowledge, as well as providing a social outlet.

Perhaps one of the best known examples of women's traditional expertise in making nutrients more available through post-harvest processing is the *nixtamalization* of maize, the staple of Mexican cuisine. This process involves soaking the corn/maize in water and limestone (calcium hydroxide) which causes the kernel skin to peel off, and releases the vitamin niacin and the amino acid tryptophan, thereby making the maize far more nutritious. The resulting maize flour (*masa*) is made into tortillas, tamales and a hot drink (*atole*) (Fitting 2011, 16). Interestingly, when maize was introduced from the New World to Europe, where it became a staple in some places, people suffered from the potentially fatal condition of pellagra. This was because they did not know that maize needed to be *nixtamalized* in order to release niacin (vitamin B3), as Mexicans had discovered centuries before. Mexican rural women grow a variety of maize, depending on the hardness and softness of the maize kernels and their suitability for different culinary purposes, particularly with regard to tortilla making. A study carried out in one region confirmed from laboratory tests the validity of women's empirical knowledge; the varieties preferred by women were "the most resistant to the local weather, were the most nutritious, and had the highest tortilla yields due to their capacity to absorb water" (Cabrera et al. 2001, 92).

Another example of women's ingenuity in post-harvest processing is burghul, a staple in Turkey, Lebanon, Syria, and Kurdistan whereby wheat is parboiled, dried and then coarsely or finely ground, depending on what it is required for: fine for salads, such as *tabouleh* and *kibbeh*, and coarse for pilafs and stews. This process gives wheat a longer shelf-life, as well as making it quicker to cook and more digestible. Lebanese food writer,

Annise Hellou (2013), describes the way in which women sort and wash the grain before parboiling it, spreading it on sheets to dry in the sun over several days; then, after it has been sent to be milled, sifting and sorting it into fine and coarse grades before storing it in bags where it will last for one year until the next harvest. Hellou notes that most burghul is now industrially processed, with the result that the traditional 'artisanal' method has largely vanished.

In the Amazon region of Venezuela, indigenous Arawakan women, in their role as the primary cultivators, provide the bulk of the household's calories from manioc grown in forest plots cleared for agriculture, and micronutrients from their homegardens, "while men spend much of their time providing protein through fishing" (Hoffmann 2003, 261). As food processors, Arawakan women invest a great deal of time and energy in detoxifying bitter manioc, a calorie-rich carbohydrate which is a staple of the diet. Manioc processing occurs every couple of months, taking a week to complete. The process involves peeling and grating the roots that are softened and fermented, then mixed with fresh manioc pulp, which is squeezed to extract the poisonous cyanide juice, with the damp dough being sifted and cooked into manioc cakes or grains (262). In a similar fashion, women throughout much of Africa detoxify cassava (also known as manioc and tapioca), which, after its introduction from the New World, eventually became a staple (along with maize) in the diet of many African countries, including Nigeria, Congo and Tanzania (McCann 2010, 51). In his fascinating book on the history of African cooking, James McCann (2010) explains how cassava is often mixed with other staple crops, such as yam, plantain, taro and sweet potato. When manioc was introduced to West Africa during the transatlantic slave trade, "women experimented with the plant's leaves for food despite awareness of the poisonous root" (Carney and Rosomoff 2009, 178). When the leaves, which are rich in protein, calcium and vitamins A and C, are pounded in a mortar and boiled with groundnuts, fish and oil, the toxic cyanide traces are eliminated, making them safe as a relish to eat with starchy staples (McCann 2010, 55).

Kimchee is central to the Korean diet, for its flavour, nutritional qualities and cultural associations. Over many centuries, Korean women developed techniques for salting, brining and fermenting many vegetables, fish and even meat. The pungency of these fermented foods, which is considered essential to any Korean meal, is the "result of a kind of 'controlled spoilage' and is a distinct characteristic of Korean cuisine"

(Kim et al. 1997, 8). Early records of kimchee making mention vegetables such as turnips, cucumbers, radish, eggplant, bamboo shoots and wild Indian mustard seasoned with black pepper, garlic and green onions (8). Once chillies were introduced from the New World, they replaced pepper, giving kimchee its characteristic fiery flavours (10). Kimchee falls into two categories: quick fermentation for early eating and long, slow fermentation for long-term storage. The preparation for the second type is complicated and many staged. Vegetables, in peak condition, are carefully sorted, thoroughly washed at least three times, then torn by hand or cut with bamboo knives and mixed with sea salt, crushed garlic and ginger, and ground chillies and other combinations of seasonings, depending upon the recipe of the season, region and family: these include fermented shrimps and fresh oysters, fermented anchovy paste, seaweed and sesame (34). The kimchee is fermented in large earthenware jars made from clay, which "effectively insulates the kimchee against the outside temperatures, cold or hot, and maintains a steady temperature inside....In traditional Korean homes, a large earthenware jar placed in the shade keeps the kimchee at an optimum temperature" (38). Traditionally, the jars are also placed underground to slowly ferment for use during the harsh Korean winter.

Like Indigenous women throughout the world, Australian Aboriginal women have traditionally been experts at detoxifying a number of plants in order to render them edible and nutritious. For instance, the nadoo plant whose high thiaminase content blocks the absorption of vitamin B, is pounded, winnowed and baked which releases the necessary vitamin. Baked cakes were made with nadoo flour and the green nadoo leaves eaten as a steamed vegetable (Pascoe 2014, 109). Other plants that are treated to detoxify them include Burrawang which is roasted, pounded and mashed before being left in water for two to three weeks to remove toxins (108). Similarly, pandanus and cycad nuts are put through "stringent sluicing and immersion treatments to remove the poisonously high alkaloid level" (107).

Women's Work with Animals

Women's relationship to food production is usually associated with plants; however, rural women also play an important role in the care and conservation of domestic livestock, which are important to their family's nutrition and economic welfare as well as providing valuable fertiliser.

In fact, historically, women have been instrumental in the development and preservation of livestock biodiversity. According to an FAO (2012) report, *Invisible Guardians: Women Manage Livestock Diversity*, "there is good reason to believe that women played a key role in the process of domesticating the major livestock species" (5) following the transition from hunter-gatherer societies to the development of settled agriculture during the Neolithic period. This is because, as male hunters had to move further afield for game, small wild animals, such as goats and sheep were brought back alive, and from this the domestication of animals began. Because women were more 'tied to hearth and home' it is likely that they played the major role in nursing, taming and raising the first livestock. Indeed, from "ethnographic observations about women nursing young animals, scientists believe that women played a major role in the taming of young stock and in bonding between humans and young animals during the early phases of domestication" (6).

The keeping of livestock is of vital importance to subsistence production, with 70% of the world's poor being livestock keepers (Gura 2008, 24), with "the world's poorest people depending on pigs, yaks, cattle, sheep, lamas, goats, chickens, camels, buffalos and other domestic animals" (*Farming Matters* 2010, 8). Of the estimated 900 million poor livestock keepers in the world, two-thirds are rural women (Chamamuto and Hall 2015). Despite the central importance of rural women's role in the care of domestic livestock, this is generally overlooked by social scientists, development planners, as well as most feminist scholars (Sachs 1996, 103). This is partly so because women's central role in food production in general is insufficiently acknowledged, but also because work with livestock tends to be associated with men. Although both men and women are involved in livestock production, there is usually a distinct gendered division of labour. Women's role in livestock management tends to involve small numbers of small animals: poultry and pigs and small ruminants, such as goats and sheep, which are largely kept for milk, eggs and wool, as well as occasional meat.[1] These small animals are free-ranging near the household, require very little land and are low-maintenance, being fed on household waste. Animals are especially important to women who have access to little or no land, because the sale of even a couple of eggs or some milk "may imply being able to buy some rice, and thus, instead of having one meal a day, a second one becomes a reality" (*Farming Matters* 2010. 8).

Men's role usually involves larger numbers of large animals, principally cattle, that are used as draught power and for meat, and that are raised on pastures and rangelands at a distance from the household (Sachs 1996). The exception to this is women's involvement in the household dairy cow which provides milk for the family. Indeed, historically rural families throughout much of what is now the 'developed' world kept one or two house-cows and poultry which were the responsibility of women who collected the eggs and milked the cow, making cheese and butter from the surplus milk. This was not only an important source of nutrition, but also a source of income for rural and peasant women, as it still is in much of the Global South. With urbanisation and the development of large-scale industrialised farming during the past century, rural women in the industrialised west have lost control of egg and milk production, a process that is increasing also in much of the newly industrialising societies of the Global South. Regarding the American situation, Sachs describes how, during the early part of the twentieth century, farm women produced most of the country's poultry. The sale of eggs by farm women to their local grocer was—in addition to trading eggs for food and dry goods—their "most common and reliable source of income" (107). However, during the twentieth century, despite women's resistance, "large-scale egg and broiler production gradually pushed women out of the poultry industry" (107). A similar process also occurred with the takeover of women's work with dairy cows, when "men, with mechanized equipment …gradually (took) over dairying and butter making …as the status and profitability of these enterprises increased" (112).

This situation is now being repeated in India under the 'white revolution' in a development named 'Operation Flood' in which native low-yield, but disease-resistant local breeds are being replaced by high-yielding European breeds, such as Holsteins and Jerseys. This development threatens the rich genetic biodiversity of local breeds that have been bred over many centuries to be resilient to the local environment and climate. In contrast, the imported cattle breeds are not adapted to the local conditions and climate, are susceptible to disease and sensitive to heat, often having to be housed in air-conditioned units (115). In the Global South, it is often women who use and develop animal breeds. Therefore, as local breeds disappear, women's role is dramatically affected in a negative way (Gura 2008, 61). For instance, the livelihoods of rural women in India are being undermined by this development; the narrow reductionist approach of the industrialised dairy industry "equate cattle only with milk, and do not consider other livestock produce – draught power, dung for fertiliser and fuel, etc." (cited, 116). A serious consequence of this development is the undermining of the livelihoods and well-being of rural women and their families. Seventy million rural households in India keep dairy animals which supply a significant source of income, as well as valuable nutrition in the form of milk, ghee, buttermilk, yoghurt and curds (GRAIN 2016). The corporatisation of dairy affects women in particular: "Investors and big dairy corporations are working hard to hijack dairy markets in India…If they succeed, it would spell economic and social disaster for millions of people" (190). The corporate dairy industry and the resulting displacement of rural women's traditional dairy industries result in wealthy urban customers being able to buy more milk; however, the rural poor consume less milk because they are unable to pay the higher prices (Sachs 1996, 115). The industrialisation of animal agriculture, such as factory farming of chickens and pigs, is rapidly increasing throughout the Global South, resulting in the transfer of control and livelihoods from small-scale women farmers to men and corporations.

Apart from providing food and money to rural households, farm animals are essential to sustainable agricultural systems. In the words of Indian ecofeminist, Vandana Shiva (2010): "The integration of livestock with farming has been the secret of sustainable agriculture. Livestock perform a critical function in the food chain by converting organic matter into a form that can be easily used by plants" (165). Indeed, so important are cattle to agriculture in India for providing fertiliser and for ploughing, that in the Deccan region there is a religious festival every April in which the cattle and their manure are celebrated/worshipped (Mazhar et al. 2007, 26). Women farmers value the manure not only for their own crops, but also as a source of income by selling any surplus. The ownership of a cow and/or goats is especially important to poor landless rural women who, for instance in the Deccan, have grazing rights for their animals on the verges of fields and to crop residues in the fields after harvest, the animals in turn benefitting the farmer by fertilising his or her fields with their manure and urine (50).

Women livestock keepers are instrumental in maintaining local livestock biodiversity that has adapted to local conditions over many generations. They prefer such breeds to the so-called improved modern breeds for a number of reasons. These include the fact that they are adapted to the local climatic conditions, are disease-resistant, tend to be small and manageable, and are low input, subsisting well on local forage. In contrast, modern breeds "have lost the behavioural traits and metabolic features necessary for accessing, utilizing and metabolizing the natural vegetation, (instead) requiring highly digestible feed" (FAO 2012). In addition, a great advantage of locally adapted livestock is that they are a 'self-reproducing asset', whereas cross-bred and 'improved' animals tend to suffer from fertility problems. Similarly, hybrid hens that are promoted by development programmes have lost their brooding-instinct, and can no longer reproduce on their own, with the result that commercial farmers have to regularly purchase new flocks (FAO 2012). Women farmers generally prefer small numbers of animals, often because they do not have the means or land to support them, but also, importantly, because they do not regard them solely in terms of being a commercial commodity. Rather, their animals serve many functions, including as an important source of nutrition, as well as a source of fibre and manure, as a buffer against environmental and financial risks and shocks, and as a potential source of income. It is therefore "the locally adapted breeds, which have often been in the family for many generations, that fit the bill, rather than production-orientated improved animals". Consequently, women "tend to like animals that are friendly, easy to handle and worry-free. They prefer such animals over those that may have high production potential, but require more time, attention and inputs" (2012).

Although my focus in this chapter is on women food producers in the Global South, it is important to note that an increasing number of women are involved in sustainable food production in the Global North. In her book examining the increase in the number of women farmers in the United States, rural sociologist Carolyn Sachs reports that in recent years there has been a phenomenal rate of increase in the number of women farmers, and that they "now comprise 30% of farm operators in the US" (Sachs et al. 2016). In the United Kingdom, almost 20% of farmers are women, an increase from 7% in 2007–2008, and with women making up 64% of graduates in agricultural studies in 2017–2018 (O'Neill et al., *The Guardian*, 16 August 2020). Male farmers who operate large-scale commodity farms concentrating on the production of oilseeds, pigs and beef cattle, and who have more motives to uphold the status quo, "adhere to a monologic masculinity (that) approaches farming as a domination of nature" (Sachs et al. 2016, 5). In contrast, most women farmers run small-scale farms with a diversity of crops including vegetables, fruits, nuts and horticulture, as well as a variety of animals, including poultry, sheep, goats and horses (8). Women farmers in the Global North, like those in the Global South, overwhelmingly avoid the use of synthetic agricultural chemicals and highly mechanised farming, both out of necessity and by choice. Sachs points out that the increasing concern about industrial agriculture and the growing interest in sustainable, organic, local food systems has provided women farmers in particular with a ready market for their high-value produce, adding that "women farmer's involvement in environmentally sustainable agriculture...parallels the greater concern that women in general have for environmental issues" (22). Furthermore, women farmers, who, in both the United States and Europe, "increasingly reject the position of farm wife" view food as more than a commodity, valuing community relationships, and promoting "local food in the context of broader international movements such as food sovereignty, slow food, and food justice" (145). Sachs and her fellow authors propose a Feminist Agrifood System Theory (FAST) as:

> a critique and an alternative to the conventional and patriarchal agricultural system. This involves gender equality for women on farms; asserting the identity of the woman farmer; accessing the resources enabling them to farm, including access to land, labor and capital; shaping new food and farming systems, and forming supportive networks between women farmers. (141)

It is important to note that many women farmers in the Global North tend to have a particular interest in rare breed animals. This is in part due to the fact that traditional breeds are under threat from the promotion of modern animals that are bred for high productivity at the expense of traits such as disease resistance and ability to adapt to a changing climate. This interest is reflected in the fact that women are the majority in rare breed trusts: in the Rare Breed Trust of Australia all six of the board members are women; in the United Kingdom's Rare Breeds Survival Trust six out of nine staff members are women; and in the Livestock Conservatory of the United States ten out of twelve staff are women.

Women throughout the world are involved in innumerable ways developing, supporting and maintaining sustainable food systems. In Australia, for instance, two women in a small rural town in the state of Victoria, concerned about "the disconnect between farmers doing great sustainable farming" and customers who wanted to buy their produce, in 2012 developed the not-for-profit Open Food Network. This growing network, which now operates in thirteen countries around the world, including France, the United Kingdom and United States, Canada, Brazil, Italy and the Philippines, supports different distribution systems, connects farmers with their communities and puts "the power of buying and selling back into the hands of the consumer and producer" (ABC, 24 May 2020). Yet, another example of women's pro-active role in ensuring healthy food for their families and communities is that of the Japanese women who started the now global movement of Community Supported Agriculture (CSA) in the 1970s in response to the mercury contamination to seafood as a result of the notorious Minamata disaster (*Farming Matters*, March 2018, 11).

Globally, women predominate in urban agricultural activities for their families' subsistence needs (Hovorka et al. 2009). In 1996 the United Nations Development Programme estimated that 800 million people worldwide were engaged in urban agriculture (5). With the rapid urbanisation of the world's population in the years since, the need for urban and peri-urban food production has even greater importance. It is estimated that by 2030, throughout the world, urban areas will triple in size, putting enormous pressure on food security as a result of the loss of peri-urban farmland (Bryce 2016). Although the erosion of peri-urban agriculture is a major problem for many countries in the Global South, it is also a serious issue in many parts of the Global North, compromising food security in a world with an increasingly precarious and unstable climate. Urban and peri-urban agriculture is of increasing importance in climate-friendly sustainable food systems. In both the Global South and North women are at the forefront of the struggle to re/claim urban land for food production. Carolyn Sachs notes that in the United States "anecdotal evidence suggests that women predominate in urban farming" (2016, 54), and, from my own observation and experience of urban community gardening, that is also the case in Australia.

Access to urban land for displaced rural communities is especially important in much of the Global South. For instance, in Africa overall 40% of urban residents engage in agriculture, producing vegetables and fruit, eggs and milk (Olivier 2016), and in West African cities it is estimated that between 60 and 100% of fresh vegetables are produced for local consumption, with the proximity to markets making urban agriculture a source of employment and livelihood as farmers, distributors, processors and marketers for the urban poor (Bryant et al. 2017,

34). However, even though urban agriculture is of great importance in contributing to food security and reducing the risk of malnutrition and even starvation, it generally receives little or no support from government, and is often made illegal. Nevertheless, despite the obstacles and prohibitions to urban gardening in many cities, women persist and find ingenious ways to grow food for the subsistence needs of their families. However, there are exceptions to such official prohibitions, such as Cape Town where the municipality acknowledges the challenges and benefits specifically to female urban food producers. With women making up the majority of about 6000 farmers on the Cape Flats, they not only benefit with regard to increased food security, but also, importantly through social benefits, including networking with other women farmers, and cultivating as groups. This social capital and support also enables them to confront patriarchal oppression and bullying from male farmers. The women farmers of the Cape Flats give away surplus produce to schools, and to those who cannot afford to buy food; in contrast the male farmers take less home than the women (20% compared to 40%), preferring to sell most of their produce (Olivier 2016).

In this chapter I have described how female small-scale farmers practise sustainable food production throughout much of the world. By applying organic agroecological methods, incorporating mixed livestock and plant agriculture, and replenishing natural resources, women farm in harmony with nature. This stands in stark contrast to the mechanised, dominating approach of masculinist industrial agriculture, with its origins in the Scientific Revolution, as described in Chapter 5. I have examined how women have historically been and continue to be promoters and protectors of biodiversity, both plant and animal, and their commitment to the preservation of seed diversity. Again, this contrasts with the modern masculinist system of industrialised agribusiness which treats food production—both plants and animals—purely as commodities, with the emphasis on maximum production for maximum profit. This reductionist, extractive approach has resulted in massive damage to the environment, people's health and great cruelty to animals in factory farms, as well as a drastic loss of both plant and animal genetic diversity. It is increasingly destroying small-scale farms, leading to the displaced populations moving to cities in search of work and where they are forced to live in squalid slums. Industrialised agriculture is based entirely on the use of fossil fuels, making it a major contributor to global warming. It has driven global environmental destruction through land-clearing and deforestation, which is also a huge contributor to global warming, as well as causing devastating loss of biodiversity, both animal and plant. Indeed, recent pandemics over the past two or three decades, including avian flu and SARS, are considered by reputable sources, such as GRAIN and IPES (Independent Panel of Experts on Sustainability), and epidemiologists who are not in thrall to agribusiness, to be caused by the massive global growth in factory farming of animals over the past few decades.

Women's food-related creativity and innovation, and their ability to improvise and make do particularly comes to the fore during times of crisis, such as the COVID-19 pandemic of 2020. As an article in an Indian publication points out, the COVID-19 pandemic is an opportunity to re-evaluate a lot of things, including recognition that women have always been the gatekeepers of health and nutrition, and "to start acknowledging the fact that women not only do the heavy lifting in most kitchens, but *they are also the repository of knowledge about sustainable and nutritious eating in good times and bad*" (Bhowmik 2020; italics added). The writer states that this acknowledgement is lacking because patriarchal society disdainfully considers the roles of caregiving and housework as menial and non-productive. She points out that "history is filled with kitchen innovations produced by women in response to spoilage, rationing or penury", such as during the Second World War and the Great Depression. She describes how in India, her own mother, during decades of upheaval in Assam, developed innovative ways of creating nutritious and delicious recipes, that have served as an inspiration to other women "creating a knowledge pool cutting across boundaries and time". A fascinating example of women's culinary innovation is how women in the eastern states of Bengal and Bihar started using poppy seeds in curries, during the time of the British Raj, when forced poppy seed cultivation limited alternatives. The result was "a stroke of ingenuity"—delicious curries, with the poppy seeds adding "taste and bulk to the curries and cooling the body in the hot, humid climate of the eastern states" (Bhowmik 2020). It is this kind of innovation and knowledge which has sustained humanity for millennia, in the process creating the world's myriad nutritious and delicious regional cuisines—in short, creating civilisation itself. It is high time that women's central role in this is fully recognised, supported and celebrated.

Note

1. However, there are exceptions to this, such as Mongolia, where women in the Gobi Desert "perform practically all activities connected with camel raising, including milking, herding, taking care of the young and milk-processing" (FAO 2012, 30).

References

Australian Broadcasting Corporation. 2020. "Violet Town Farmers Find Global Success During Coronavirus Pandemic." https://www.abc.net.au/news/2020-05-22.

Biscotti, Nello, and Andrea Pieroni. 2015. "The Hidden Mediterranean Diet: Wild Vegetables Traditionally Gathered and Consumed in the Gargano Area, Apulia, SE Italy." *Acta Societatis Botanicorum Poloniae* 84 (3): 327–338.

Bhowmik, Tanushree. 2020. "Even Before Coronavirus, Indian Women Were Passing Down Secrets of Quarantine Cooking." April 11. https://scroll.in/article/print/958792.

Braudel, Fernand. 2002. *Memory and the Mediterranean*. New York: Vintage.

Bryant, Christopher, Jorge Pena Diaz, Bernard Keraita, Frank Lohrberg, and Makato Yokohari. 2017. "Urban Agriculture from a Global Perspective." In *Urban Agriculture Europe*, edited by Frank Lohrberg, Lilli Licka, Lionella Scazzosi, and Alex Timpe. Berlin: Jovis Verlag.

Bryce, Emma. 2016. "Growing Mega-Cities Will Displace Vast Tracts of Farmland by 2030." *The Guardian*, 27 December.

Cabrera, Isidro Rimiarachin, Emma Zapata Martelo, and Veronica Vazquez Garcia. 2001. "Gender, Rural Households, and Biodiversity in Native Mexico." *Agriculture and Human Values* 18: 85-93.

Carney, Judith A. 2001. *Black Rice: The African Origins of Rice Cultivation in the Americas*. Cambridge, MA and London: Harvard University Press.

Carney, Judith A., and Richard Nicholas Rosomoff. 2009. *In the Shadow of Slavery: Africa's Botanical Legacy in the Atlantic World*. Berkeley: University of California Press.

Chamamuto, Nicola, J. C., and Stephen Hall. 2015. "Gender Equality, Resilience to Climate Change and the Design of Livestock Projects for Rural Livelihoods." *Gender and Development* 23 (3) (November): 515.

Daniggelis, Ephrosine. 2003. "Women and 'Wild' Foods: Nutrition and Household Security Among Rai and Sherpa Forager-Farmers in Eastern Nepal." In *Women and Plants: Gender Relations in Biodiversity Management and Conservation*, edited by Patricia L. Howard, 83–97. London and New York: Zed Books.

Dogan, Yunus. 2012. "Traditionally Used Wild Edible Greens in the Agean Region of Turkey." *Acta Societatis Botanicorum Poloniae* 81 (14): 329–342.

Ertug, Fusun. 2003. "Gendering the Tradition of Plant Gathering in Central Anatolia (Turkey)." In *Women and Plants: Gender Relations in Biodiversity Management and Conservation*, edited by Patricia L. Howard, 183–194. London and New York: Zed Books.

Ertug, Fusun. 2009. "Wild Plant Foods: Routine Dietary Supplements or Famine Foods?" In *From Foragers to Farmers*, edited by Andrew S. Fairbairn and Ehud Weiss, 64-72. Oxford and Oakville: Oxbow Books.

FAO. 2012. *Invisible Guardians: Women Manage Livestock Diversity*. Author Ilse Kohler-Rollefson, Food and Agriculture Organisation Animal Production and Health Paper No. 174, Rome, Italy.

Farming Matters. 2018. "What is Community Supported Agriculture?" ILEIA, March.

Farming Matters. 2010. "Going for More Livestock: The Benefits of Livestock." ILEIA, March.

Fitting, Elizabeth. 2011. *The Struggle for Maize: Campesinos, Workers, and Transgenic Corn in the Mexican Countryside*. Durham, NC: Duke University Press.

GRAIN. 2016. *The Great Climate Robbery: How the Food System Drives Climate Change and What We Can Do About It*. New Delhi and Melbourne: GRAIN, Daraja Press, Tulika Books and Spinifex Press.

Gray, Patience. 1987. *Honey from a Weed: Fasting and Feasting in Tuscany, Catalonia, the Cyclades and Apulia*. London: Papermac.

Greenberg, Laurie S. Z. 2003. "Women in the Garden and Kitchen: The Role of Cuisine in the Conservation of Traditional House Lot Crops Among Yucatec Mayan Immigrants." In *Women and Plants: Gender Relations in Biodiversity Management and Conservation*, edited by Patricia L. Howard, 51–62. London and New York: Zed Books.

Gura, Susanne. 2008. "Livestock Breeding and Public Policies: Impact on Smallholders and Their Breeds." Consultancy report to the League for Pastoral Peoples and Endogenous Livestock Development. www.pastoralpeoples.org.

Handy, Jim, and Carla Fehr. 2011. "Drawing Forth the Force That Slumbered in Peasants' Arms: The Economist, High Agriculture and Selling Capitalism." In *Food Sovereignty: Reconnecting Food, Nature and Community*, edited by Hannah Wittman, Annette Aurelie Desmarais, and Nettie Wiebe, 45–59. Cape Town, Dakar, Nairobi and Oxford: Pambazuka Press.

Hoffmann, Shirley. 2003 "Arawakan Women and the Erosion of Traditional Food Production in Amazonas Venezuela." In *Women and Plants: Gender Relations in Biodiversity Management and Conservation*, edited by Patricia L. Howard, 258–269. London and New York: Zed Books.

Hovorka, Alice, Mary Njenga, and Henk de Zeeuw, eds. 2009. *Women Feeding Cities: Mainstreaming Gender in Urban Agriculture and Food Security*. Rugby: Practical Action Publications.

Howard, Patricia L. 2003a. "The Major Importance of Minor Resources: Women and Plant Biodiversity." International Institute for Environment and Development, Gatekeeper Series No. 112, January. Published by International institute for Environment and Development.

Howard, Patricia L. 2003b. "Women and the Plant World: An Exploration." In *Women and Plants: Gender Relations in Biodiversity Management and Conservation*, edited by Patricia L. Howard, 1–48. London and New York: Zed Books.

Howard, Patricia L. 2006. "Gender Bias in Ethnology: Propositions and Evidence of Distorted Science and Promises of a Brighter Future." Conference: Distinguished Economic Botanist Lecture, Kew Royal Botanical Gardens, UK, November.

Howard, Patricia L. 2012. "Human Resilience in the Face of Biodiversity Tipping Points and Regional Scales." In *Addressing Tipping Points for a Precarious Future*, edited by Tim O'Riordan and Tim Lenten, 104–126. Oxford: Oxford University Press.

Hellou, Anise. 2013. *Levant: Recipes and Memories from the Middle East*. London: Harper Collins.

Ibnouf, Fatima Osman. 2009. "The Role of Women in Providing Household Security in Sudan: Implications for Reducing Hunger and Malnutrition." *Journal of International Women's Studies* 10 (4): 144–167.

Kaneva-Johnson, Maria. 1999. *The Melting Pot: Balkan Food and Cooking*. Totnes, Devon: Prospect Books.

Kim Man-Jo, Lee Kyou-Tae, and Lee O. Young. 1997. *The Kimchee Cookbook: Fiery Flavors and Cultural History of Korea's National Dish*. Singapore: Periplus.

Kochilas, Diane. 2014. *Ikaria: Lessons on Food, Life, and Longevity from the Greek Island Where People Forget to Die*. Rodale Publishing, New York.

Lopez-Alzina, Diana Gabriela, and Patricia L. Howard. 2012. "The Structure, Composition, and Functions of Homegardens: Focus on the Yucatan Peninsula." *Etnoecoligica* 9 (1): 17-41.

Mazhar, Farhad, Daniel Buckles, P. V. Satheesh, and Farida Akhter. 2007. *Food Sovereignty and Uncultivated Biodiversity in South Asia*. New Delhi: Academic Foundation.

McCann, James M. 2010. *Stirring the Pot: A History of African Cuisine*. London: Hurst and Company.

Nazarea, Virginia D. 1998. *Cultural Memory and Biodiversity*. Tuscon: The University of Arizona Press.

Oliveri, Valentina. 2015. *The Flavours of Calabria*. San Vendemiano: SIME Books

Olivier, David W. 2016. "Uprooting Patriarchy: Gender and Urban Agriculture on South Africa's Cape Flats." The Conversation, 10 March.

O'Neill, Holly, Tim Lewis, and Sirin Kale. 2020. "We Are Stewards of the Land: The Rise of Female Farmers." *The Guardian*, 16 August.

Pascoe, Bruce. 2014. *Dark Emu: Black Seeds: Agriculture or Accident?* Broome, WA: Magabala Books.

Pieroni, Andrea, and Renate Soukand. 2017. "The Disappearing Wild Food and Medicinal Plant Knowledge in a Few Mountain Villages in North-East Albania." *Journal of Applied Botany and Food Quality* 90: 58-67.

Pionetti, Carine. 2005. *Sowing Autonomy: Gender and Seed Politics in Semi-arid India*. London: The International Institute for Environment and Development.

Pimbert, Michel. 2018. "Global Status of Agroecology: A Perspective on Current Practices, Potential and Challenges." *Economic and Political Weekly* 53 (41): 52–57. https://www.epw.in/journal/2018/4/1review-environment-and-development/global-status-agroecology.html/.

Sachs, Carolyn. 1996. *Gendered Fields: Rural Women, Agriculture and Environment*. Boulder, CO: Westview Press.

Sachs, Carolyn E., Mary E. Barbercheck, Kathryn J. Brasier, Nancy Ellen Kiernan, and Anna Rachel Terman. 2016. *The Rise of Women Farmers and Sustainable Agriculture*. Iowa City: University of Iowa Press.

Satheesh, P. V. 2014. "Cultivating Biodiversity; Peasant Women in India." *Farming Matters*, March.

Shiva, Vandana. 2010. *Staying Alive: Women, Ecology and Survival in India*. Melbourne, VIC: Spinifex Press.

Shiva, Vandana. 2016. *Who Really Feeds the World?: The Failures of Agribusiness and the Promise of Agroecology*. Berkeley, CA: North Atlantic Books.

Tittonell, Pablo. 2015. "Safe-Guarding Diversity in Home Gardens." *Farming Matters*, December.

CHAPTER 9

Conclusion

The subject of food, despite being a generally taken-for-granted subject, is in fact deeply political and also highly gendered. Women especially have a complex relationship to food, in large part related to their role as nurturers. Women's identity is bound up with food in both negative and positive ways—on the one hand, eating disorders, guilt, sometimes seen as a source of daily drudgery, and on the other hand, pride and pleasure in being creative and sharing with others, as well as a sense of empowerment and authority. While acknowledging women's often ambivalent relationship to food issues, my focus in this book has been on the many positive aspects of women's relationship to food. In Chapter 1, I described women's historical role in the development and maintenance of the world's innumerable healthy, delicious and sustainable regional cuisines. And in the final chapter I described the role of women today in many parts of the world as producers and growers of food, including their promotion and defence of biodiversity, which is massively under threat from industrial agriculture and global warming, of which industrial food systems are a major contributor.

It has been my intention in this book to describe women's generally unacknowledged historical role in the world's sustainable food systems, how women have for millennia developed an immense reservoir of plant knowledge, of both agricultural and wild plants involving seed selection

V. A. Swinbank, *Women's Food Matters*,
https://doi.org/10.1007/978-3-030-70396-7_9

and preservation. It is rarely acknowledged, except by some ethnobotanists, that the serious issue of the preservation of genetic diversity is inextricably linked to the domestic sphere and the kitchen. Drawing on the important work of ethnobotanists, such as Patricia Howard, I have described how women's traditional botanical knowledge is intimately related to food preparation, and the importance of the interconnectedness between women's traditional role in growing and gathering plants and cooking. This has resulted in the world's innumerable delicious and nutritionally sound regional cuisines. I assert that this intergenerational life-sustaining female knowledge must be acknowledged and indeed celebrated. This is truly something that the world's female population has cause to be deeply proud of.

Women's so-called 'unproductive' work in the domestic sphere is not valued, or is at least hugely undervalued, and even rendered invisible in male supremacist culture, even though in fact such work is highly productive and life-sustaining. The devaluing of women's reproductive work that is the foundation of society has been especially acute since the development of industrialised capitalist economies starting with English Acts of Enclosure of the commons as discussed in Chapter 6. I have argued that it is very important that women's role in sustainable food systems—from biodiversity to culinary diversity—is fully acknowledged, including by women themselves, and not allow it to be denied or appropriated by men. A prime example of this is the high status accorded to men's (especially professional) cooking vis-à-vis women's domestic cooking as I have analysed in both Chapters 3 and 4.

It is of serious concern that both genetic and culinary diversity and the associated knowledge and skills, which have been handed down intergenerationally, largely through the female line as discussed in Chapter 5, are increasingly under threat from the many processes described in this book. These processes have particularly accelerated during the last half-century: industrialised food systems from production to preparation; free trade deals; the dominance of supermarkets providing a false illusion of (packaged) diversity; the displacement globally of small-scale farmers by mega-farms and land grabbing and, increasingly, climate chaos, which are all discussed in Chapter 7. An additional issue of great concern is the loss of women's culinary knowledge due to a number of factors, including the double work burden. Carole Counihan (1988), for instance, describes how young Italian women, despite valuing the relative independence and self-esteem that paid work brings, experience a sense of frustration that, due to time constraints, they are unable to maintain the standard of their mother's and grandmother's cooking, as well as losing the associated influence. Similarly, in southern Italy young women no longer have the wild plant knowledge of their foremothers, and consequently lose the accompanying status and authority (Pieroni 2003). Patricia Howard (2003) reports that when girls from Indigenous communities go to school the education system devalorises women's traditional subsistence knowledge and practices, which disinclines them to carry them on (38).

One important subject that, for lack of space, I have not addressed in this book is that of food entitlement. Despite producing much of the world's food, especially in the Global South, and throughout the world being almost entirely responsible for food preparation for their families, women are often disproportionately discriminated against with regard to adequate nutrition. The right to adequate nutrition is enshrined as a basic human right by the United Nations Universal Declaration of Human Rights (1948). However, as Indian economist Amartya Sen (1984) states, there is ample evidence that shows that within families the world-over there is a distinct sex bias against women and children regarding food allocation, with the privileging of adult males, especially regarding high-quality foods, such as protein. This is despite the fact that in many societies, women's workload and therefore energy expenditure is greater than that of men (Bina Agarwal 1988). This has been exacerbated in recent years with the increasing feminisation of agriculture due to men's migration to cities in search of work, leaving women with an even heavier agricultural workload. Culturally imposed food taboos which are common in India and parts of Africa are especially targeted at women, leading to malnutrition, particularly during pregnancy. Furthermore, women in general tend to police their own food intake as a result of sexist socialisation which expects that women be self-denying (Swinbank 2008).

In this book I have described how women are central to sustainable and healthy food systems that stand in stark contrast to the destructiveness of masculinist industrialised agribusiness. This has especially come to the fore with the COVID-19 pandemic which, along with other pandemics, such as SARS and Avian flu, is a direct consequence of the corporate industrial food system which maximises production and profit above human health, animal welfare and environmental concerns. In Chapter 7 I discussed how the outbreak of Ebola in West Africa was caused by the destruction of native forests for industrial-scale palm oil production. It is now widely accepted by epidemiologists that pandemics in recent years, including SADS (swine acute diarrhoea syndrome) and SARS (sudden acute respiratory syndrome) are a result of the massive increase of industrial animal production, particularly poultry and pigs. This global growth, referred to as the Livestock Revolution, plus the increasing interface of urban and peri-urban development is increasing the risk of deadly viruses, caused by natural habitat destruction, particularly of forest ecosystems (Wallace 2016; Davis 2005). Despite industrial food systems being the main cause of pandemics, it is small-scale free-range farms that are targeted by governments and government-backed corporations for closure. However, with recognition that factory farming of animals is the cause of pandemics, the solution to preventing future outbreaks must involve the phasing out of industrialised animal production, with a return to traditional non-industrial methods of farming, drawing on the knowledge of peasant farmers, especially women peasant farmers around the world.

If nothing else, the COVID-19 pandemic should have taught us to have a far greater respect and appreciation of nature, and to realise that when nature is abused, as it has been for several centuries, at least since the Industrial Revolution, it will eventually turn on us. Central to the abuse of, and indeed war, on nature has been the global expansion of industrial agriculture since the Second World War—deforestation, the destruction of biodiversity, the total dependence on fossil fuels, chemicals, pesticides, herbicides and GMO, and most seriously, global warming largely caused by these factors. Although the making/foundations of this disaster have a long history starting with the Scientific and Industrial Revolutions, as I described in Chapter 5, it is during the last few decades of the twentieth century and the first two of the twenty-first that the consequences of the extractive, unsustainable food system have come to a head. In his classic book, *Small Is Beautiful: A Study of Economics as If People Mattered*, published in 1973, Eric Schumacher analyses how capitalist's pursuit of profit and conviction that 'bigger is better' is at the expense of the environment and social welfare. However, tragically, the principles of Schumacher's analysis were completely overturned in the 1980s by the rise of neoliberal capitalism, whereby 'greed is good' became the dominant mantra, and which has led to the current disastrous environmental and social problems, the most recent of which is global pandemics, most recently COVID-19.

The subject of food is vast and open-ended. The issues I have examined in this book are ongoing and, if anything, are likely to intensify in the future—the increasing industrialisation of the food system; the accelerating loss of biodiversity, food knowledge and skills; and increasing inequity in access to adequate nutrition, especially for women and children, as a result of globalised corporate food control, and the negative effects on food supply caused by global warming. As the dominant paradigm of industrial production attempts to increase its control over the global food system, the struggle against this and the sense of urgency in promoting and preserving sustainable food systems will also intensify.

In "*Women's Food Matters*" I have examined/analysed the complex and many-faceted relationship women have with food production and preparation, which have been and are being affected by industrialisation and commodification of the food system. I have examined the loss of skills and knowledge that have resulted from these processes, a loss that I believe will become even greater as the globalised forces of free trade agreements intensify corporate control, standardisation and homogenisation of the world's food supply, undermining both cultural diversity and biodiversity. However, there is growing awareness and resistance to this, and an increasing urgency in saving the richness and diversity of the world's myriad food cultures from such standardisation. By acknowledging and celebrating women's vital role in the creation and preservation of biodiversity and related culinary cultures, it is my hope that this book makes some contribution to that effort.

References

Agarwal, Bina. 1988. "Neither Sustenance nor Sustainability: Agricultural Strategies, Ecological Degradation and Indian Women in Poverty." In *Structures of Patriarchy, the State, the Community and the Household*, edited by Bina Agarwal. New Delhi: Kali for Women; London: Zed.

Counihan, Carole M. 1988. "Female Identity, Food, and Power in Contemporary Florence." *Anthropological Quarterly* 61 (2): 51–61.

Davis, Mike. 2005. *The Monster at Our Door: The Global Threat of Avian Flu.* New York: The New Press.

Howard, Patricia, L. 2003 "Women and the Plant World: An Exploration." In *Women and Plants: Gender Relations in Biodiversity Management and Conservation*, edited by Patricia L. Howard, 1–48. London and New York: Zed Books.

Pieroni, Andrea. 2003. Wild Food Plants and Aberesh Women in Lucania, Southern Italy. In *Women and Plants: Gender Relations in Biodiversity*

Management and Conservation, edited by Patricia L. Howard. London: Zed Books.

Sen, Amartya. 1984. "Family and Food: Sex Bias in Poverty." In *Resources, Values and Development*. Oxford: Blackwell.

Schumacher, E.F. 1974. *Small Is Beautiful: A Study of Economics as If People Mattered*. London: ABACUS.

Swinbank, Vicki A. 2008. "The Sexual Politics of Food: A Radical Feminist Analysis of Food Production, Preparation and Consumption." Unpublished PhD, University of Melbourne.

Wallace, Rob. 2016. "Did Neoliberalizing West African Forests Produce a New Niche for Ebola?" In *Big Farms Make Big Flu: Dispatches on Infectious Disease, Agribusiness, and the Nature of Science*, 324–335. New York: Monthly Review Press.

Index

V. A. Swinbank, *Women's Food Matters*,
https://doi.org/10.1007/978-3-030-70396-7

The manufacturer's authorised representative in the EU is Springer Nature Customer Service Centre GmbH, Europaplatz 3, 69115 Heidelberg, Germany. If you have any concerns regarding our products, please contact ProductSafety@springernature.com

Printed and bound by CPI Group (UK) Ltd, Croydon, CR0 4YY
30/09/2026
02225702-0005